DATE DUE

NO 01 '04			
NO 12 '04			
MAR 26 '08			
OCT 23 '08			
NOV (			

GAYLORD PRINTED IN U.S.A.

Praise for NIGHT HAS A THOUSAND EYES

"In *Night Has a Thousand Eyes*, Art Upgren not only reminds us of the simple pleasures of star gazing, he reconnects us with the generations of those who have dared to ask how it all works. Upgren's book thus carves a path along which we can reacquaint ourselves with the Universe."

—Dr. Neil deGrasse Tyson, Director, Hayden Planetarium, New York, New York

"I wish I had *Night Has a Thousand Eyes* when I began observing the night sky more than 35 years ago. Clearly written and easy to understand, [this] book will point you to the stars in a way that should make them lasting friends."

—David H. Levy, author of *Sharing the Sky* and *Impact Jupiter*

"Anyone who has ever looked at the sky in wonder, be it from the dark of the country or from the light-polluted city, will find much in this book to help in the understanding of celestial phenomena visible to the naked eye, and their influence on our art and literature. This is a highly entertaining and informative book."

—C.A. Murray, formerly Head of the Division of Astrometry and Galactic Astronomy at the Royal Greenwich Observatory

"What better introduction to astronomy could the culture-conscious layman or beginner student in this field find than *Night Has a Thousand Eyes*? Throughout the book one is impressed how humor with relevant connotations leading to the more scientific explanations can be an effective teacher's tool for holding attention and clinching serious facts."

—Dorrit Hoffleit, Senior Research Scientist, Department of Astronomy, Yale University

"A breathtaking guide to the immense universe . . . invaluable to the layman."

—William Manchester, author of *American Caesar: Douglas MacArthur, 1880–1964* and *The Last Lion: Biography of Winston Churchill*

"What a jewel! This book has a thousand delights for the reader—astronomy, science, the sky, poetry, even history and lore, all wrapped up into a delightful bundle. It sparkles with the author's knowledge and enthusiasm."

—Tim Hunter, Past-President, International Dark-Sky Association, Inc. (IDA)

"With lilting prose, *Night Has a Thousand Eyes* weaves history, poetry, and literature into this primer on naked-eye astronomy and appreciation for the wonders of the dark night sky. As an introduction to sky lore, astronomy, and intellectual history, it in turn sees the "thousand eyes" through many eyes."

—Richard Berendzen, Professor of Astronomy, American University

"A well-written and refreshing approach to the appreciation of the night sky."

—Roy Garstaing, Professor Emeritus of Astrophysical and Planetary Sciences, University of Colorado at Boulder

"Seamlessly mixing poetry with astronomy and lore, Arthur Upgren provides an engaging introduction to our common heritage—the night sky."

—Joan D. Hedrick, author of *Harriet Beecher Stowe: A Life*

"Emerson observed, 150 years ago, that because of modern improvements, so-called, 'the man in the street does not know a star in the sky . . . the whole bright calendar of the year is without a dial in his mind.' The situation now is, if anything, worse. We must therefore be enormously grateful to Arthur Upgren for his valiant attempt to remedy this state of affairs. Upgren here provides the perfect introduction to the wonders of the night skies, and an eloquent plea to curb light pollution so we can actually see those skies with our own eyes. You don't need a telescope. You don't even need binoculars. But you do need this wonderful book."

—Robert D. Richardson, Jr., author of *Henry Thoreau: A Life of The Mind* and *Emerson: The Mind on Fire*

NIGHT HAS A THOUSAND EYES

A Naked-Eye Guide to the Sky, Its Science, and Lore

PERSEUS BOOKS
Reading, Massachusetts

NIGHT HAS A THOUSAND EYES

A Naked-Eye Guide to the Sky, Its Science, and Lore

ARTHUR UPGREN

Library of Congress Cataloging-in-Publication Data

On file

ISBN 0-306-45790-3

Perseus Books is a member of the Perseus Books Group.

3 4 5 6 7 8 9 10   02 01 00 99

Printed in the United States of America

To my father, my first teacher.
In so many ways, this is his book.

Contents

Foreword

I wish I had *Night Has a Thousand Eyes* when I began observing the night sky more than 35 years ago. On a warm evening in August of 1960, I had nothing but a small telescope. With nothing but me and the stars, I had no idea where to start. So with scope in hand, I turned to the brightest "star" in the sky. That star, it turned out, was Jupiter—its four moons and two dark bands gave away the presence of the biggest planet in the solar system.

On that night I had no idea that the distant planet would someday be a central focus of my life. In July 1994, a comet I helped discover, Comet Shoemaker–Levy 9, struck Jupiter, producing the largest explosion ever seen in the solar system. For me, it was the culmination of many happy and interesting years under the night sky. These years were spent first learning the constellations one by one, and then finding, through a telescope, the deeper wonders that were out there.

With the help of Arthur Upgren's book, you too now have the chance to go out on a starlit evening and make your own acquaintance with the stars. The author has a vast experience with the sky and with teaching it to others. Clearly written and easy to understand, his book will point you to the stars in a way that should make them lasting friends.

David H. Levy
Vail, Arizona

NIGHT HAS A THOUSAND EYES

A Naked-Eye Guide to the Sky, Its Science, and Lore

Introduction

The night has a thousand eyes, And the day but one;
Yet the light of the bright world dies with the dying sun.
The mind has a thousand eyes, And the heart but one;
Yet the light of a whole life dies when love is done.

—Francis William Bourdillon

The night has a thousand eyes and more; they mark the sky in an almost changeless manner, moving only in concert with the sky itself, appearing to spin around us every day. Since the dawn of recorded history, the same stars in nearly the same array have shone down on all people, giving all of our civilizations a common heritage.

From many years of teaching introductory and descriptive astronomy, I've learned that most students are eager to become familiar with the brighter stars and constellations and the planets as well. I always try to include evening sessions with my courses, in which the stars and constellations form the show. These are not laboratory sessions, but times simply to enjoy the sky and its contents. I have also participated in programs held at our national and state parks to point out to visitors the salient features of the night sky. Tourists visiting these areas are becoming increasingly aware that a dark night sky forms an integral part of our natural environment to be experienced, enjoyed, and protected from the excesses of air pollution and the ever brighter man-made illumination known as light pollution.

This book is a response to the enthusiasm shown by many of my former students who still delight in identifying the features of the night sky, especially when they have the opportunity to see it in its full splendor away from the lights and haze of the city. Nothing but the naked eye is needed to appreciate its beauty and wonder. The legacy of the night sky with its constellations and their curious-sounding names, names used by Homer and Plato and Dante and Shakespeare, binds us to our own heritage. The names are still in use today and form the scientific nomenclature in longest continuous use of any; four thousand years have passed since the best and brightest were given the names by which we know them today. They are worth knowing, and once known they are not easily forgotten. It is my hope that their recognition will pass from the agrarian society to the urban society intact and that they always will be seen as a worldwide preserve for us all.

The first part of the book deals with the stars and their recognition. They are, to a first approximation, immutable. The wanderers among them comprise the subject of the second part. The Sun and Moon and the five bright planets are each a unique world whose physical natures and motions I will describe. They form the constantly changing cast seen against the starry backdrop. Throughout I have shown many ways in which the sky inspires the arts, literature, and the other sciences. In these associations, too, the night has a thousand eyes. Today the telescope transforms the science of astronomy, but that was not always so. The formation of our heliocentric worldview and much other science came about when the naked eye was our only tool. As individuals, we repeat our collective lore and appreciate the night sky. This is a book devoted to that worldview.

Arthur Upgren
Middletown, Connecticut
Sanibel Island, Florida

Part I

The Starry Skies

CHAPTER 1

Getting Acquainted with the Changing Sky

If a man would be alone, let him look at the stars. The rays that come from those heavenly worlds, will separate between him and what he touches. One might think the atmosphere was made transparent with this design, to give man, in the heavenly bodies, the perpetual presence of the sublime. Seen in the streets of cities, how great they are! If the stars should appear one night in a thousand years, how would men believe and adore, and preserve for many generations the remembrance of the city of God.

—RALPH WALDO EMERSON
Nature

On a clear night, away from the lights and haze of city and suburbs, there seems to be no end of stars. So many are visible that the familiar brighter constellations are almost lost among them. If the Moon is not around, the faint, luminous band of the Milky Way can often be seen arching overhead.

Back in the city, or even in the suburbs, the sky doesn't appear this way. Buildings and trees may obscure much of the sky, but even if they do not, the glare from bright streetlights gives the sky a pale, washed-out appearance. The very brightest stars and planets are plainly visible, but they aren't impressive. Like a Christmas tree in sunlight, even with its lights on, the city sky is bright and the stars do not stand out in relief. The starry sky goes too often unnoticed in town, where only a handful of stars can be seen down a street or over a neighbor's garage. Furthermore, stars and constellations change positions in the sky from one season to the next in a way that may make their recognition and identification appear difficult. As with the weather, the stars and the patterns they make have long been the special province of the farmer, the shepherd, and the seafarer. This is unfortunate since many of us—city and country dweller alike—react with curiosity and interest when we are able to see a dark sky in all its glory.

The night sky is most approachable in summer when evening temperatures are balmy and travel to countryside, mountains, and seashore is at a maximum. Becoming familiar with the stars and constellations of summer is a helpful starting point for appreciating the slowly changing pageant of the skies as they move throughout the year. The slow westward motion of the stars becomes evident as we enter into the crisp clear nights of autumn. Later, Orion and the brilliant winter stellar groups appear to rise until they dominate the evening sky at Christmastime. In spring the westernmost of the summer stars return in the east, as old friends. Using no equipment other than our eyes, we will examine the skies of each season in turn. We will also describe some special events occurring in the summer, including the famous Perseid meteor shower, which fills the sky with shooting stars in mid-August to the delight of vacationers everywhere.

All the while, we are seeing the stars of the past. Stars are so distant that it takes years, even centuries, for their light to reach us. One of the brightest summer stars, Vega, is seen as it appeared 25 years ago, and another, Deneb, is seen as it appeared back in late Roman times, or the early Middle Ages.

After describing the stars and their motions, we will direct our attention to those bright wanderers that seem to move freely about the background of the so-called fixed stars, named for their apparent rigidity within their respective constellations. These are the readily visible planets—Mercury, Venus, Mars, Jupiter, and Saturn—and once familiar, they are very easy to identify throughout the year.

These five planets, along with the Sun and Moon, have given the number seven its special significance down through the ages and their names to each of the seven days of the week. The inclusion of the Sun among the nighttime wanderers may at first seem odd, but in fact, our forebears have been acutely aware for many centuries that the only difference between our central star and its planetary retinue is one of brilliance. At dusk some set of stars is always present, forming a background for the Sun, whose light obscures their visibility to the naked eye, scattered as it is by the Earth's atmosphere.

Differences in longitude do not affect the appearance of the sky, because those east–west differences amount to changes in time only; however, changes in latitude impose the most striking shifts in the appearance of the stars we see. Astronomers and navigators long before Columbus were aware of this fact. Our North American viewpoint is centered on the region between 39 and 42 degrees north latitude. An unusually large number of the most populous regions of the United States and Canada are found within these latitudes, including Boston, New York, Philadelphia, Washington, Baltimore, Pittsburgh, Cleveland, Cincinnati, Detroit, Chicago, Saint Louis, Kansas City, and Denver. An extension to Toronto, Buffalo, and Milwaukee to the north and to Richmond, Louisville, and San Francisco to the south could be made in almost every case without loss of accuracy. Even Montreal, Ottawa, Minneapolis, Portland, and Seattle are not very far off the mark to the north, nor are Atlanta, Memphis, and Los Angeles to the south. In Europe, Madrid, Rome, Athens, and Istanbul lie in this region of latitude, as do Tokyo, Seoul, and Beijing in Asia.

A change in latitude larger than this does indeed alter the aspect of the heavens. In present-day North America, we are witnessing the greatest regular mass migration in history over different latitudes as, each winter, millions of northerners make their way south to Florida and farther to the Caribbean Islands. The deep southern sky is probably the most wondrous and luminous of any region, and many fortunate travelers will find the Southern Cross and other brilliant "new" stars, in addition to our more familiar northern stars in new and different aspects. Our description of the changes will show those who have the opportunity to see, as ancient navigators have seen, the effect on their view of the heavens made by a simple flight to the south.

Observers living in the British Isles view the sky from more northerly latitudes. Britain extends from 50 degrees north latitude at Land's End to 60 degrees in the Shetland Islands. If we look at eastern North America in the same range of latitude, we find Labrador. It is easy for Americans and Canadians to forget the offset of Britain and Europe to the north of them. How do the British Isles manage to support a highly industrialized society of 60 million people, while Labrador is home to only about 30 thousand, scrabbling amongst bare rocks on land supporting little vegetation and almost no agriculture? As has been known since Ben Franklin discovered it, the Gulf Stream makes much of the difference, thwarting the southward progress of pack ice and the frigid weather that can accompany it.

In 11,500 B.C. or thereabouts, the glaciers that had dominated northern North America, Europe, and Asia for so long began a retreat that was completed three thousand years later, as far as we know. With their retreat, ocean levels began to rise as less of the world's water remained locked up in frozen form. The warming and melting continued, and sometime after 7000 B.C. the English Channel broke through and connected the Atlantic Ocean with the North Sea. The Thames was no longer a tributary of the Rhine flowing into a diminished North Sea, and Britain became an island, with a guaranteed maritime climate free of most extremes in temperature.

Thus London lies at latitude 51½ degrees north, about the latitude of Calgary, Alberta, and almost 11 degrees north of New York, whose latitude matches that of Madrid and Naples. The skies of Britain differ significantly from the skies over the great cities of North America for this reason. The latitudes of Paris, Berlin, and Vienna are close to that of London. Days and nights vary more widely with the seasons. The Sun sets around teatime in the winter, and lighting-up time is past midevening in the summer. From London north the sky never gets fully dark in June and July. Yet the seasons are mild. Whenever the night sky is clear and relatively free of city lights, the circumpolar region, the part of the sky that never appears to set, is larger as seen from London north. The bright stars Capella and Deneb and much of the constellation of Perseus are circumpolar, and Vega is nearly so. But in exchange, the United Kingdom and northern Europe lose the southern regions of the sky; for the British, much of Sagittarius and Scorpius are never visible, and the tail of Canis Major and the star Fomalhaut creep along at their petty pace near the southern horizon and don't stay up for long. Significant differences between the two latitude zones will be mentioned as appropriate throughout this book. As the inhabitants of Albion flee to southern resorts in the wintertime and settle near the Mediterranean Sea, they find themselves at latitudes comparable to New York and Washington or, if they venture farther south, to the Canary Islands, on a par with the cities of southern Florida.

Here we cover the sky with the naked eye. Telescopes and binoculars add greatly to one's appreciation of the heavens, and a number of books are available that discuss the sights they can provide. But the first acquaintance with the sky is best made of the *whole* sky with our own unaided eyes.

Chapter 2
In Praise of Twilight

The curfew tolls the knell of parting day,
The lowing herd winds slowly o'er the lea,
The ploughman homeward plods his weary way,
And leaves the world to darkness and to me.

This familiar opening quatrain from Thomas Gray's *Elegy Written in a Country Churchyard* addresses what is arguably the most vivid and dramatic sky phenomenon we see: the transformation of day into night. These lines could describe the fading light of an overcast autumnal English afternoon, or they could apply equally well to a clear summer evening just before sunset, when the shadows lengthen and stretch across lawns and over hedge and furrow. The changes occurring over the next hour or two are as dramatic as anything we observe in nature. In that time interval the light intensity of the sky overhead diminishes to about one four hundred thousandth its brightness at sunset.

That we too often ignore the changes of dusk is a mark of its regularity, predictability, and frequency; only the diurnal certainty of this pageant detracts from its spectacle. Dusk has for rivals only the reverse process we call dawn and the more sudden darkness of a total eclipse of the Sun by the Moon. Such rare events as total solar eclipses appear at one location only once in several centuries, and astronomers crisscross the Earth chasing after them whenever and wherever they may occur. In contrast, the eclipse of the Sun by the Earth (for that, after all, is what "night" means) is commonplace; the full panoply of twilight occurs every clear night.

Fortunately our eyes are adaptable to great changes in light. We are able to handle effortlessly the transition from the solar glare of daylight to a darkness in which we can see a single candle at a distance of ten miles. The first step in that transition occurs shortly before sunset. The clear sky overhead at the zenith is somewhat fainter and darker than when the Sun was higher in the sky, and the Sun itself has a noticeable reddish tinge. This is because red light with its longer wavelength penetrates directly through the atmosphere, but the shorter-wave blue light gets scattered by the same air. The principal components of our atmosphere, the nitrogen and oxygen molecules that form all but one percent of this atmosphere, are much smaller than the wavelengths of visual light. They scatter incoming light in a very predictable manner called Rayleigh scattering. It affects the shorter wavelengths in the blue and violet regions of the visual spectrum much more than the longer wavelengths of the red end. The unevenness between colors accounts for not only the blue of the daytime sky, but also the increasing redness (and faintness) of the Sun, Moon, and stars as they approach the horizon, where they must shine through much more of our atmosphere than they do aloft.

Although Rayleigh scattering is always present, haze is not; it varies with atmospheric conditions. Haze is composed of aerosols, or low-level particulate matter of a larger size arising from a variety of sources (fine dust, water droplets, and matter due to air pollution). Haze scatters colors evenly. A third, minor cause of optical extinction by the atmosphere is the molecular absorption brought about by carbon dioxide, ozone, and water vapor.

* * *

After the Sun has set below the horizon, it is worth noting a few of the transient phenomena of the darkening sky. Ten or twenty minutes after sunset on a very clear evening the sky takes on several different hues. The cerulean blue of the daytime sky quickly gives way in the west to a fiery rose-pink, while the blue lingers overhead. But it is toward the east that we see the most marked changes. Pink is present there too in a more subdued form; however, along with the deepening of the sky, a band of purple rises from the eastern horizon to meet and blend into the brighter pink above. To understand the reason for the deep purple color, we must note that the atmosphere some miles above our heads is still in the sunshine at this point. (If this were not the case, the transition from day to night would take place in an instant, as it does on the Moon, where there is no air to scatter the incoming sunlight.)

After sunset, the sky along the eastern horizon appears purple because it no longer receives sunlight at any height above the ground, being now hidden from the Sun by the rim of the Earth. This purple layer is fully within the Earth's shadow and is the first visible harbinger of true night.

On evenings when clouds cover much of the heavens, most of the preceding description is not valid. Clouds to the west will alter the pink and purple of the eastern sky to a blue-gray. The dark band of the Earth's shadow may still be visible, but it will not be distinctly purple. Even if a twilight sky contains only small clouds, the fading sunlight will have wrought some striking changes in a matter of minutes. Before sundown, all clouds are white (or leaden gray if sheets of stratus clouds cover most of the sky). The white ones are of two basic types. The rounded, puffy, fair-weather summer-day clouds are cumulus; cirrus clouds are thin and feathery with ragged edges. The basic difference between them is one of altitude: The cirroform clouds, formed at high altitudes, are composed of ice crystals; the cumulus type, found in the warmer air at lower levels, contain water droplets. Many variations of each and combinations of the two are often present.

Just after sunset, any nearby hills are still swathed in the solar glare, but soon they too fall dark. The clouds that may be present above remain brilliantly illuminated for a while, but in time they will also darken. Most striking are the sunsets with both cirrus and cumulus clouds present in the sky. The cumulus clouds change quickly from a rosy luminescence to gray as the sunlight withdraws, and the higher cirrus do the same later on. As each cloud level is plunged into darkness in its turn, a curious transformation occurs. The clouds to the west become dark against the still-bright clear sky beyond, while their counterparts to the east remain brighter than the background sky. The high reflectiveness of clouds is at no time more apparent than at sunset, and these conditions give rise to the most spectacular sunsets of all, with colors ranging from fiery reds and pinks to subdued mixtures of orange and gray. This striking transformation may escape the notice of most astronomers, amateur and professional alike. But it did not escape that of Claude Monet, who used the interplay of light and shade in some of his series paintings. He and some of his fellow Impressionists were keenly aware of the directional interdependence of the relative light intensities of clouds and clear air.

After sunset, clouds appear dark against the still-bright western sky over Boston.

After sunset, clouds in the eastern sky of Miami Beach appear bright against the darker sky since they are better refectors of light than the clear air beyond.

Approximately half an hour after sunset, a number of changes occur. The Sun is now about six degrees (or about 12 of its own angular diameters) below the horizon, and it is roughly at this point that motorists turn on their headlights, sensing this step toward the dark.

We are passing rapidly from bright to deep twilight, and we can now see the first stars. In the eastern sky the purple edge of night becomes indistinct, merging with the heavens above it. The sky overhead retains its color but the darker landscape becomes a mixture of grays in the fading glow. This is because the human eye loses its ability to detect colors even before darkness renders it unable to see shades of relatively bright and dark light intensity. The crepuscular gloom gives rise to the nocturnal insecurity many of us feel when we cannot see distinctly. Shaded areas become peopled with trolls and vampires and other creatures of legend, as we cross that fine line between our world and one in which "they" take over. The atavistic sense of unease and menace that may come upon us at twilight derives in part from the ages when primitive people faced wolves and other predatory creatures that could function in the darkness better than humans.

It is no wonder, then, that we look upward toward Moon and stars for reassurance. The first stars have indeed become visible by now but they are the very devil to find. They are not easily seen at this time of twilight because only the brightest handful are visible, and they are lost in the immensity of the empty celestial hemisphere. Later on, when they twinkle vividly in the black sky, their fainter companions are much more visible because the bright ones act as signposts pointing the way to them.

The remaining story is one of ever-deepening nightfall. The last vestiges of twilight are still quite visible in the western heavens as the constellations fill in around the brighter stars. The Sun will still make its presence noticed for another hour or more at our latitudes; only when it is a full 18 degrees below the horizon at the end of astronomical twilight will the last of its rays vanish, allowing night—full darkness—to take over. Even in the absence of the Moon and bright planets, the clear night sky is never fully dark, black though it may sometimes appear, since the stars remain to cast some light.

The open sky between the stars also emits a faint glow, which derives from four separate component sources. The first of these, aloft in the upper reaches of our atmosphere, is the airglow caused by the ambient excitation of upper-atmosphere atoms and molecules by energetic particles streaming from the Sun. The airglow is green, but it is usually so faint that human eyesight cannot come close to detecting it. In cases of extreme solar activity, this bombardment causes the phenomenon of the aurora. When it is seen surrounding the North Magnetic Pole in northern Canada, it is known as the aurora borealis, or northern lights. Its southern counterpart, the aurora australis, is not frequently noticed because few live far enough south to see it. When the aurora is bright to the point that color can be detected, it is usually green because it is an excess of airglow, of ions and electrons bombarding the upper air typically 50 to 100 miles above the ground. The color is ascribed to two lines in the green portion of the spectrum that arise from the so-called forbidden transitions. These lines cannot under most conditions be reproduced in the laboratory and were once thought not to occur anywhere—hence the designation forbidden. But now we know that they are found in the near-vacuum conditions in space. The common green lines are due to doubly ionized oxygen (atoms of oxygen with two electrons missing), and some red lines, due to singly ionized nitrogen, are occasionally also seen in a bright auroral display. They appear at a higher level and if seen behind a green patch, blend with it to form a spot of yellow-orange.

Beyond the Earth we find the second source of glow: zodiacal light, the sunlight reflected by the ever-present interplanetary dust particles that are continually enriched from the residue of disintegrating comets and other debris. Beyond our solar system lie the third and fourth sources: the integrated light from stars too faint to be seen by the naked eye as well as the light from the nebulae and diffuse matter in our Milky Way galaxy. Both are additional contributors to the background light, although in smaller amounts. The light from these four sources taken together, emanating from an area of sky equal to the size of the full Moon, contributes less than the light from a single star at the threshold of naked-eye visibility. This is faint indeed but still noticeably brighter than the pitch-black conditions found in a cave or well.

Somewhere in the northern sky the Big Dipper now takes its place as the most familiar of all stellar groups. On every clear night of the year the Big Dipper is a reassuring portent that all is as it should be in an orderly universe. Beyond it the sky appears disordered and confusing to the uninitiated. The Big Dipper is an American term; in Great Britain its seven bright stars are known as the Plough. There and elsewhere in Europe this figure is also called Charles's Wain, the wagon or chariot, usually, of Charlemagne.

The welter of stars visible on a good clear night, especially a moonless night, makes their identification seem hopeless at first glance. We can sympathize with James McNeill Whistler, the

American painter, who may have felt this way when he was invited to step outside to view the stars on a rare clear night in London. He demurred, stating that "there are far too many of them and they are so very poorly arranged."

Orientation is not difficult, however, if we make a few observations using the Big Dipper as our starting point. Dusk is a good time to do this, since dusk is when the dipper comes into its own. In summer it is seen to the northwest about midway between the horizon and the zenith overhead.

Four conditions can interfere with our observations. First and most obvious, the sky can be overcast. Second and least disruptive, a full or nearly full Moon will alter the appearance of the heavens during the final stages of dusk as it begins to outshine the western twilight at about the point when the brighter constellations become visible. The third condition occurs whenever too much of the sky is obscured by trees and buildings.

The last is the most pervasive disruption of all. The full beauty of a dark sky is hidden from all but about one American in twenty. The rapid increase in the number and brightness of streetlights has brought about a parallel increase in the brightness of the night sky, called skyglow. Of the 6000 stars that the average human eye might be able to see in the entire sky, about 2500 are visible at any one time on a clear, moonless night. These are reduced to only a few hundred even in a moderately streetlit suburb. In recent years this problem of light pollution has been addressed by the lighting industry, and today light fixtures are available that reduce upward illumination and reduce glare to the side. This lighting preserves safety and even enhances it while at the same time greatly reducing skyglow. As municipalities and institutions become aware of the savings in money and energy brought about by the use of newer energy-efficient lamps directing the light only downward where it is needed, the lovely phenomenon of twilight will once again become the heritage of a majority of Americans, and we will again restore John Milton's "Circling canopie of Night's extended shade."

Chapter 3
The Big Dipper and the Constellations of the Northern Sky

It is always necessary to remember that constellation in medieval language seldom means, as with us, a permanent pattern of stars. It usually means a temporary state of their relative positions.

—C. S. Lewis
The Discarded Image

In the United States and in Europe, summer sunsets occur quite late in the evening, especially in areas that observe daylight saving time. Twilight lingers and often the sky is not fully dark until after ten o'clock. In the summertime the brightest part of the sky at dusk is always found toward the northwest horizon. With this in mind, one can determine the other compass points easily. If the sky is completely dark when first observed, another reference point must be used.

Among all of the constellations, or star groups, in the sky, the Big Dipper, or the Plough as it is customarily known in Great Britain, is by far the best known. Its high degree of recognition arises from three of its properties. It does not rise and set when seen from our midnortherly latitudes, but is visible all night, every clear night of the year. This is because it is *circumpolar*, a term used to describe the portion of the northern sky and the stars located in it that are so close to the pole that they do not set. They do not sink low enough to reach the horizon as the entire sky appears to rotate around the Earth once every 24 hours, reflecting the Earth's own rotational motion about its axis. The motion makes the Sun and Moon seem to move westward along with every other object in the sky and accounts for their rising in the east and setting in the west. The sky must rotate about two points or pivots, which must necessarily be located directly above the Earth's north and south poles, the points at which the Earth's axis intersects its surface. The two celestial poles form the extension of that axis to the apparent celestial sphere. From the United States we see the point in the sky lying directly above the North Pole, about halfway up from the northern horizon to the point directly overhead, called the zenith.

The second reason for the high recognition of the Big Dipper is that the seven stars that form it are among the brightest in the sky. Thus it is easily seen, even under hazy or moonlit conditions, or whenever the sky is not fully dark. These stars form part of a much larger constellation known as Ursa Major, the great bear, but the rest of this huge constellation is composed of much fainter stars and is not conspicuous. Finally, the two bright stars forming the edge of the bowl of the dipper on the side away from the handle are known to many people as the pointers because they point toward another bright star about as far away as the dipper is long. This lone star is Polaris, the north star, or pole star. Polaris is located very near the actual pole of the sky, the point directly above the Earth's North Pole and about which the sky appears to rotate. It thus indicates true north and is a very useful star to recognize. It will also show up in almost the very same place every clear night of the year. (Polaris actually does move slightly, but this tiny motion goes unnoticed without a telescope.) The dipper and the north star, whose presence will reassure those fearful of the dark, can serve as starting points for familiarizing oneself with the other stars and constellations.

Polaris is part of the Little Dipper, or, more properly, Ursa Minor, the little bear. Like its larger neighbor, this dipper is also formed by seven stars, four of which are very faint and invisible except on the clearest and darkest of nights. In town, the Little

Dipper appears as only three stars. In addition to Polaris at the end of the handle, only the two stars forming the edge of the bowl away from the handle (in the analogous position to the pointers of the big dipper) are visible. The other four stars all lie in between the two and the pole star. This makes for an unimpressive group, especially when compared to its illustrious larger companion.

Polaris is bright but it is not among the very brightest stars in the sky. Almost 50 others are brighter. Since it may not be easily distinguished from other bright stars to the beginning observer, the Big Dipper forms a more useful first feature in the sky to identify. On summer evenings, as we noted in Chapter 2, it is located in the northwest, about halfway between the horizon and the zenith. Thus, if any twilight is still visible, the Big Dipper will be found in the same direction and above it. The handle will be pointed upward and somewhat to the left of the bowl, which may be fairly low in the sky, especially late in the summer season. Once you have found it, try to find Polaris and mark its location. Polaris is probably the most important star in the sky for one to be able to find, since along with an immediate determination of true north, its nearly unvarying position serves as the surest reference point from which to find and locate other stars and star groups. "As constant as the North Star" is an oft-quoted remark with good reason.

Now it is time to look for these stargroups in the sky. As the evening wears on, the sky rotates counterclockwise around Polaris, and the Big Dipper settles down toward the horizon. If the night is at all clear, its seven stars and the bright three of the Little Dipper should be readily visible. The figure on page 22 shows how the pointers of the Big Dipper locate Polaris, which is always due north, throughout the year.

Since they are always with us, the dippers make for a sensible example of the lengths of lines or arcs in the sky. An entire circle composes 360 degrees in angular measure, and from the figure on page 23 we can see that the Big Dipper is about 25 degrees in length and that the two pointers are 5 degrees apart. Inexperienced observers may sometimes revert to inches or feet to indicate the distances between stars, but these are *linear* units

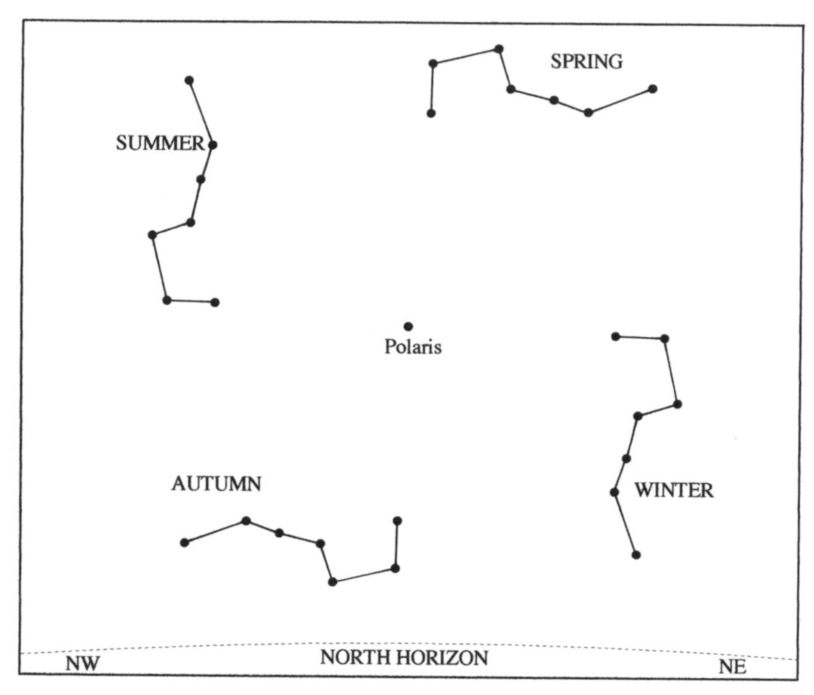

The positions of the Big Dipper at midevening at four times of the year.

of distance. In order to make sense of them in the sky, we would have to give a distance at which so many inches would look to be so many degrees. It is much more convenient to communicate these angular distances in terms of angular measure, and with a little experience, anyone can master their measure.

The Big Dipper itself is properly called an asterism, a name for an easily noticeable group of stars that does not make up a full constellation. The Little Dipper is both since its set of seven stars is nearly all one can see of Ursa Minor. This difference has been of significance since about 1920 when constellations were first rigorously defined as areas of the sky. Earlier they were star-figures, variously defined from one culture to another.

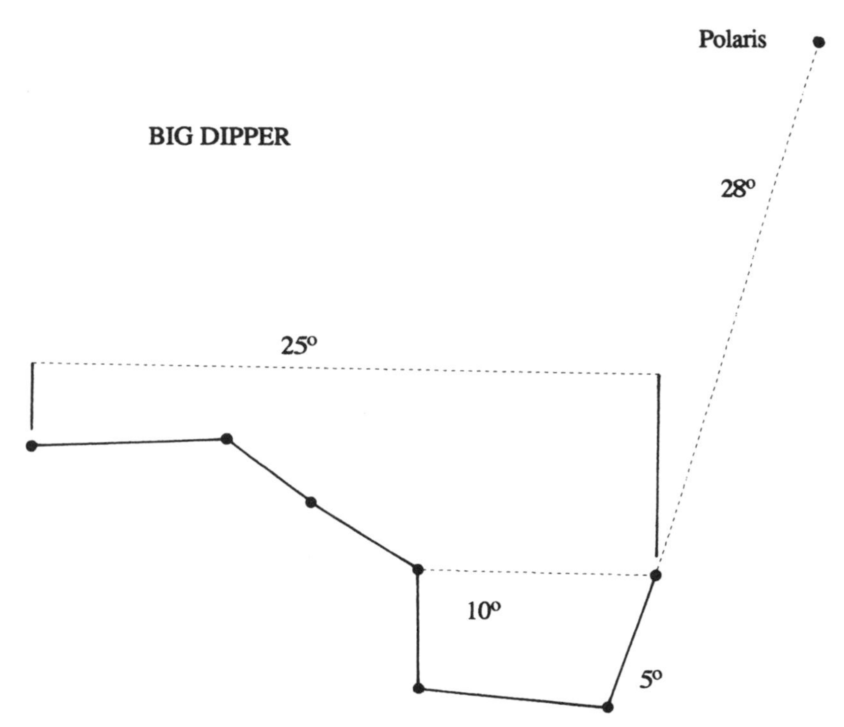

The angular separations of the stars in the Big Dipper. For comparison, the width of a hand at arm's length is about ten degrees and of one finger at arm's length, about two degrees.

A pair of stars within the Big Dipper has long been recognized as a test for good eyesight in many cultures. The brighter star is Mizar, the star at the bend of the handle and second from the end of it. The other is Alcor, a faint star just alongside and above Mizar, about a third of the Moon's apparent diameter away.

The description of Ursa Major and Ursa Minor would not be complete without mention of the fact that, as noted earlier, Ursa Major is a vast constellation and the Big Dipper comprises only the hindquarters and the tail of the giant bear. Ursa Minor is by comparison only a cub, comprising the Little Dipper and not much else, but it too is pictured with a tail that terminates at

Polaris. Earthly bears rarely, if ever, have tails. The origin of this bit of heavenly artistic license is unknown, but it is surely of very ancient origin. For the last few millennia at least, celestial bears have had tails.

* * *

After the dippers and the north star, the easiest stars to locate are a group of five that form a wide W or M, depending on the time of night and year at which it is seen. To find this group, look east of the pole star; it appears about as far from Polaris to the east as the Big Dipper is to the west. The stars are bright, about the same brightness as Polaris and the stars of the dipper. On summer evenings, the five stars form a wide W lying on its side, and they will always be on the opposite side of the pole from the Big Dipper and equidistant from it. They are also a circumpolar group; that is, they never set as seen from the midlatitudes of North America and Europe. They form the major part of the constellation known as Cassiopeia, a mythological queen of Ethiopia. The five brightest stars forming the W-shaped figure are often portrayed as a chair or throne upon which the queen is seated.

The five stars of the W of Cassiopeia and the seven of the Big Dipper form two of the most easily recognized star groups of all. Together they form a kind of giant symmetrical balance, being on opposite sides of the fulcrum of the sky marked by Polaris and equidistant from it. They play a very large role in the oral traditions of the Native Americans of Canada and the northern United States inasmuch as they dominate the entire northern sky. The lore of the Athabascan tribes of Alberta and surrounding regions, for example, centers on just this equivalence between the W and the dipper. From their homelands, the two figures are visible every clear night of the year. Small wonder that we begin with them.

* * *

It becomes readily apparent as we see the arrangements of stars in the sky that very few of them resemble the people and animals for which they were named. Much imagination went into their designation, and many people have suspected that each society attributed to asterisms, or constellations, a preconceived set of objects to be honored in the heavens. Historians suspect now that the names of present-day constellations derive mostly from the cultures surrounding the Tigris and Euphrates river basins of Mesopotamia probably late in the third millennium before the Christian Era. Animals account for the largest number of them and the ones portrayed correspond to the fauna of that area. Neither the crocodile or hippopotamus of Egypt nor the elephant or tiger of India are found in the sky, as should be expected if the star-figures of these peoples had prevailed. Even the presence of the two bears in the northern heavens has been used as evidence favoring the Mesopotamian and Assyrian origin since bears are found only to the north, in the foothills of the Caucasus Mountains. This interpretation lost some of its plausibility when it was discovered that the dippers represented bears in many different cultures widely separated in space and time (for reasons that are not at all clear). To be sure, the deep southern sky inaccessible to these cultures, and some gaps between the larger, brighter star-figures, are now filled with constellations of much more recent origin. But 48 of the 88 constellations now recognized, including all of the best-known ones, are of ancient origin. It is widely postulated that the names spread from the Mesopotamian societies to the Minoan and Mycenaean civilizations surrounding the Aegean Sea and later to the Greeks and Romans and on to us. Our technologically demanding society has divided the entire sky into 88 constellations with boundaries. Thus every point in the heavens falls into one or another of them, much as all of the land area of the Earth except Antarctica is now demarcated into states and nations with exact boundaries, but up until this century the originals were figures incorporating only the brighter stars.

The reason for the preoccupation with animals is not known, but it may stem from the same impulse that led to the creation of the treasured cave drawings at Lascaux and Chauvet in southern France some 15,000 years earlier, where animals also predominate.

Cassiopeia is one of at least six constellations connected by a single mythological tale, known as the Perseus legend, which will be covered in the section describing these star groups. Most of the others are not circumpolar but are prominent in the autumn sky. The one other circumpolar constellation of this group is Cepheus, the king, named for Cassiopeia's husband. He is much fainter and is quite outshone by his queen.

* * *

The northern circumpolar region contains two other groups worth mention here. One is a loose, winding aggregate known as Draco, the dragon, and the other is the only far northern constellation of more modern extraction intended to fill in a nearly blank area of sky. This is Camelopardus, the giraffe, and it is conspicuous only by the absence of any readily visible stars. To find it, you need only look for the large region close to Polaris with nothing in it. While on a dark night a few stars may be barely visible, for our purposes we shall quickly pass over this poor giraffe.

Cepheus and Draco, unlike the giraffe, are significant constellations shown on star maps, but not only are they composed mostly of fainter stars than the two bears and Cassiopeia, they also do not form memorable or easily recognized patterns (see the figure on page 27). Out under the dark sky of the country, the king and the dragon are easily seen; however, back in the city or the suburbs, only the brightest star or two is visible in either of them. One of our primary purposes is to focus on the stars and star-figures that can be easily found and recognized back home in the city as well. For the great majority of Americans, the boreal sky will appear to include the seven stars forming the Big Dipper, the three at the ends of the Little Dipper, and the five that give Cassiopeia her characteristic W-shaped figure.

The star in Draco worth attention is not very bright, and when under the pole (that is, when it is below the Little Dipper in the sky) it may not be visible in the city. It can be seen most of the year midway between the two bright stars in the bowl of the Little Dipper and Mizar (along with Alcor), the star at the bend of the handle of the Big Dipper.

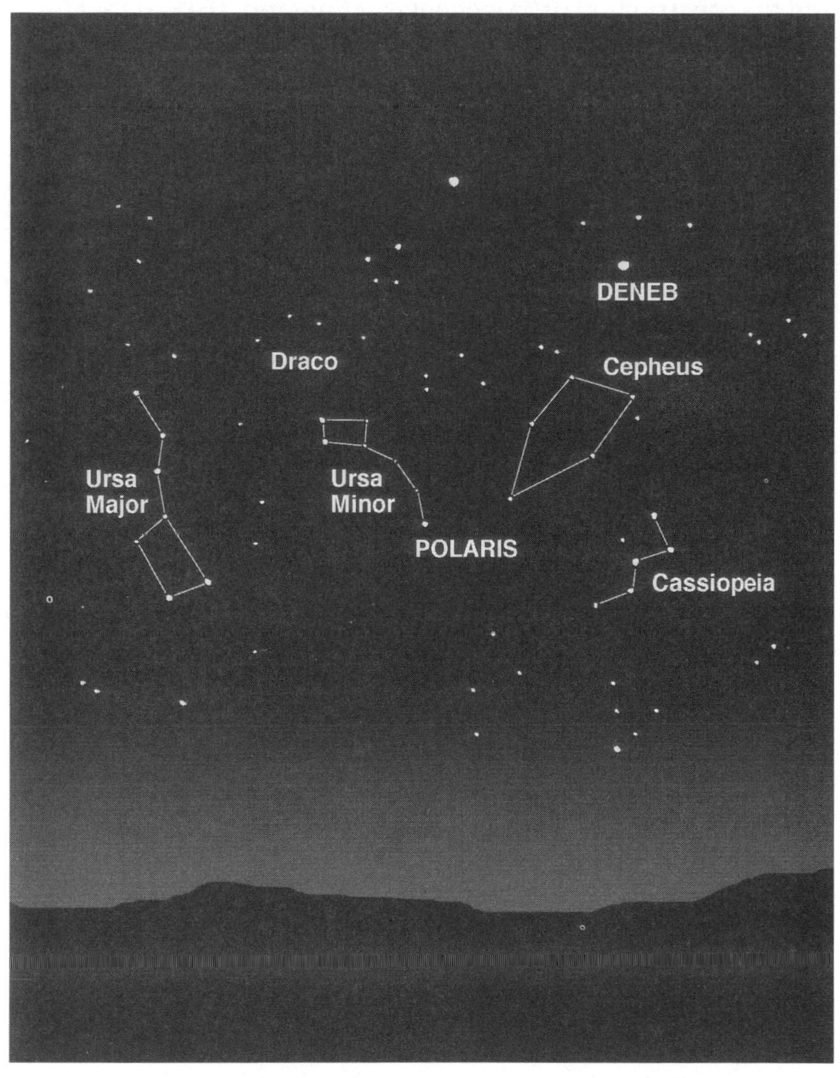

The north circumpolar region as it appears in midsummer. The Big Dipper is seen to the left of Polaris and Vega appears near the zenith.

This star, known as Thuban, is near the end of Draco's long, winding tail. Thuban was the pole, or north, star about 4500 years ago, at the time of the erection of the great pyramids of Egypt and the construction of Stonehenge in England. Back at that time, Thuban was located about where Polaris is now. If we look away from Polaris in the opposite direction from Thuban and an equal distance away, we come to the only bright star in Cepheus. This star, known as Alderamin, will be our north, or pole, star about 5000 years from now.

A huge arc can be imagined passing from Thuban through Polaris to Alderamin. The arc shows the change in the orientation of the Earth's axis over 10,000 years and is part of a giant circle in the sky. The pole describes an entire circle, of which this arc is a part, in 25,800 years. This motion is known as the precession of

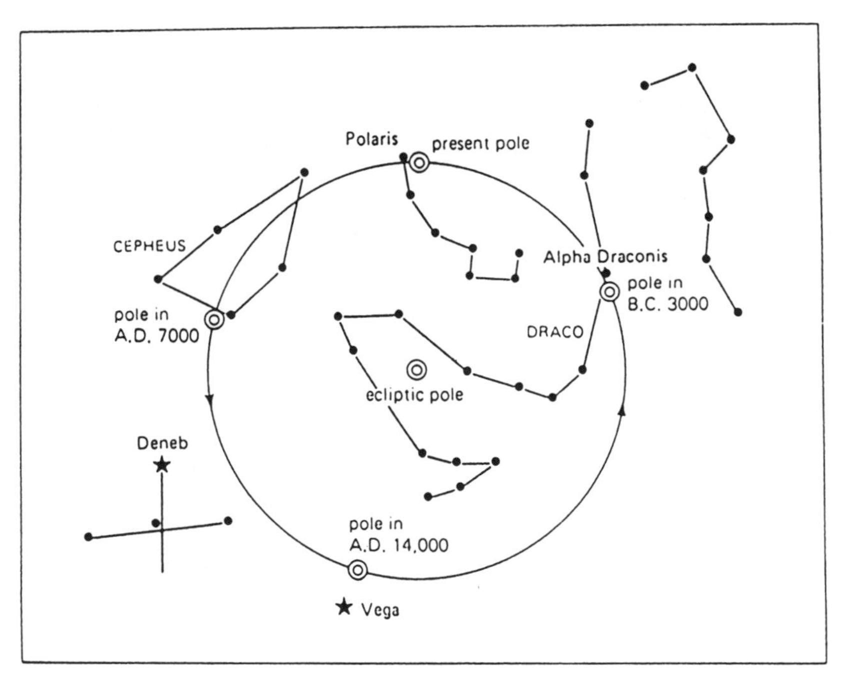

The positions of the North Celestial Pole in the sky showing the locations of past and future pole stars. Alpha Draconis is also known as Thuban.

the equinoxes, or, more commonly and simply, precession. The circle and the background constellations appear in the figure on page 28, along with pole stars past, present, and future.

Precession was first discovered by Hipparchus, one of the great astronomers of antiquity, in the second century B.C. in Alexandria, Egypt. He discovered it while comparing positions of many stars that he had determined with those of a star catalog made a few centuries earlier and now lost to us. It was a remarkable achievement, given the degree of precision necessary to detect the precessional motion among the stars.

It remained for Isaac Newton to discover and explain the cause of the precession as result of the gravitational pull of the Moon and the Sun on the equatorial bulge of the Earth. Newton realized that precession is identical to the motion of a spinning top or gyroscope as its axis starts to lean away from a vertical orientation. The Earth's gravity wants to pull it over, and it responds by precessing to the side, just as the Earth's axis moves sideways. Many aspects of the sky change slowly over the centuries due to precession, and we will discuss these changes as we view the stars of each season in turn.

Look back to Thuban now and notice its relation to the two dippers, the two bears. When Thuban marked the North Pole in the sky, the bears wheeled around it in much tighter circles than they now circle Polaris. Their association with the north was more striking in classical times than is the case today. By the same token, Cassiopeia and Cepheus were more distant from the pole then than now, and from midlatitudes they rose and set when Thuban marked the pole. They were naturally more firmly associated with the other star groups of the Perseus legend that now grace the autumn sky.

The night sky, studded as it is with stars, resembles the celestial sphere that it has often been made out to be. Many civilizations portrayed it as a round ceiling to which the stars were affixed. Only the Sun and Moon and the planets visible to the naked eye were seen to move against the starry background. This concept led to a two-sphere model of the universe, with the earth as the small sphere in the center. Although we now know otherwise, the model is still useful in such activities as surveying and celestial navigation. Anyone can easily imagine the night sky as a large sphere almost infinite in size, whereas the daytime and twilight skies are not as easy to picture in this way. If we imagine the universe to include only the solar system, the seven moving bodies that would remain could hardly produce the same effect as the stars do. Omar Khayyam's "inverted bowl we call the sky" needs stars to give it shape.

CHAPTER 4

The Summer Sky and the System of Stellar Magnitudes

Canst thou bind the sweet influences of Pleiades or loose the bands of Orion? Canst thou bring forth Mazzaroth in his season? or canst thou guide Arcturus with his sons?

—JOB 38: 31–32

Up to now we have been concentrating on the northern sky because the stars in this area are always above the horizon. They quickly become old friends, always visible whenever the night sky is clear. But any glance at the rest of the heavens, at any time of year, will reveal one very significant fact: None of the brightest stars are in the north. No matter what the month or season, the brightest stars are found in the rest of the sky—the large part that rises and sets and that can be seen only during certain times of night and certain seasons of the year.

The "rest of the sky" is indeed that great area characterized by its continual change produced by rising and setting. Any portion of it is sometimes visible and sometimes below the horizon (determined solely by the time of day or night and the time of the year), and the Sun, Moon, and planets are always found in it as are most of the stars. It is most commonly thought of as being divided into four regions named after the four seasons. Each region is highest and best seen during the evening hours of its corresponding season. Thus the stars of summer dominate the zenith and surrounding regions during the summer months, and we continue our celestial tour with them.

Let us proceed directly to Arcturus, one of the two brightest stars in the summer sky and indeed one of the very brightest in the entire sky. If we consider the three stars forming the tail of the bear as an arc, and we extend this arc in the direction away from the bowl, we come about one dipper-length away to the brilliant yellowish star Arcturus. Since this is the brightest star in the part of the sky that is highest in the springtime, it has long been associated in many cultures with spring planting.

Arcturus is included among the summer stars because it remains so conspicuous throughout the summer. It is found in the constellation Boötes, a ploughman sometimes pictured driving a plough, the English appellation for the seven bright stars of the bear. The rest of Boötes is rather faint and forms a thin, kite-shaped figure extending upward from Arcturus toward the zenith. It looks nothing like a man, and it is likely that no one ever thought it did. But people and animals were probably felt to be more deserving of an eternal place in the celestial sphere than kites and dippers, so a man it became. The position of Arcturus relative to the stars of the dipper is more important to people learning the constellations than is its position within its own stellar group because few of the other stars in Boötes can be seen in the city. Until one is very familiar with the network of brighter stars all around the sky and throughout the year, the location of the entirety of any but the brightest constellations is not an easy task.

Arcturus has been a favorite of almost all societies. It is one of the few celestial objects mentioned in the Bible. The name derives from the same root word as does *arctus* or *arctos*, words meaning "bear," associating Arcturus with Ursa Major, the great bear lying just to the north. The Big Dipper's curved handle is nearby, and a common name for both star and dipper in ancient times is understandable. In Italy the star is called Arturo, pointing out the connection to the name Arthur and the great medieval king of that name. It is even alleged that Arthur, the eldest son of Henry VII of England, the Prince of Wales and first husband of Catherine of Aragon, was named for this star. His premature death led Catherine into the first, and perhaps most fateful, of the marriages of his younger brother, Henry VIII. The identification of the far north with bears is probably a result of the very northern sky being seen as the home of the celestial bears. Hence they give the Arctic, and ipso facto the Antarctic, their names as well.

Arcturus figures once more in our history, in an event of this century. The Century of Progress exhibition was a world's fair held at Chicago in 1933 and 1934, and among the new technological wonders to be displayed there was the photoelectric cell. To demonstrate its features most dramatically, a photocell was installed on the great refracting telescope of the University of Chicago, located at the Yerkes Observatory about 80 miles northwest of the city at Lake Geneva, Wisconsin. On the opening day of the exposition, the telescope was aimed toward Arcturus, and at the proper time, the starlight shone on the photocell, producing an electric current that opened the gates to the fair. Arcturus was chosen because it was believed at the time to lie at a distance of about 40 light-years. The light opening the fair was thought to have left the bright star 40 years earlier, at the time of the previous world's fair in Chicago, the Columbian Exposition of 1892–93. Astronomers have since recalculated the distance and found that the light rays that opened the fair had actually left Arcturus only about 37 years earlier, just before the turn of the century.

The distances to stars are quite properly astronomical. Astronomers have invented long units of distance in order to avoid large numbers; the best known of these is the light-year. This is the distance traveled by a light beam in a year, at its speed of about 186,000 miles, or 300,000 kilometers, per second, and amounts to just under six million million miles, or ten million million kilometers.

* * *

If we extend our imaginary arc from the dipper past Arcturus by a dipper length once again, we come to another very bright star, Spica, not the equal of Arcturus but still outshining the dipper stars. Spica is a bluish star and lies in the constellation Virgo. None of the other stars in this celestial region are at all conspicuous and Spica, even more than Arcturus, stands alone. Especially in late summer, Spica will be seen low in the southwestern sky where the haze is greater and the fainter stars are less visible. Unless a bright planet appears nearby, these two stars, Arcturus and Spica, shine by themselves unchallenged in the whole western portion of the heavens.

High in the sky all summer long lie three other bright stars that compete with Arcturus and Spica. They rise in the evenings of late spring, moving higher and higher until they are nearly overhead and dominate the entire sky as the summer ripens to its full. Brightest of all is Vega, near the zenith all summer long. It is easy to find; it is the brightest thing at any height in the eastern heavens, and it is Arcturus's only stellar rival. Any equal of these two must be a planet and must appear only in the lower southern sky. Somewhat toward the northeast of Vega is a fairly bright star, and a little farther to the southeast is another. The first is Deneb and the second is Altair. Although each of these three bright stars lies in a separate constellation, the triangle they form is the single most distinctive feature of the overhead sky all throughout the mid- and late summer. This Summer Triangle, as it is widely called, is paramount to any further familiarization with the fainter stars. All three of these stars appear blue in color. But compare Vega with Arcturus and you should see one of the most distinct color differences between any two stars in the sky.

Glance again at these three stars. Their triangle will dominate the western sky almost up to Christmas, and by April and May they will be back to shine in the east. Thus, for three seasons of the year they command attention.

Vega and its two companions, Deneb and Altair, represent a wide range of stellar types. Two factors determine the brightness, or magnitude, of a star as it appears in the sky: its intrinsic luminosity and its distance from our solar system. The intrinsic luminosities of stars vary enormously, with the brightest outshin-

The midsummer sky seen toward the south, showing the Summer Triangle.

ing the faintest by more than a million to one. In the case of the Summer Triangle, we know that two of its stars appear among the brightest in the sky because they are close to us as stars go, whereas the third, Deneb, is one of the truly intrinsically brilliant stars in our whole Milky Way galaxy. Altair and Vega are not faint stars—far from it. Altair is 10 times as bright as the Sun and Vega is about 55 times as luminous. But neither is among the great celestial beacons of our galaxy. The nearest star to the Sun is Alpha Centauri, just over four light-years away, and Altair and Vega are among the next closest conspicuous objects. We know them to be about 17 and 25 light-years away, respectively. Thus they appear not as they are today, but as they were that many years in the past. Any inhabitants of a planet orbiting Altair (if there are any) might, for example, be able to observe the 1989 dismantling of the Berlin Wall and the collapse of the Soviet empire in the year 2006, whereas the folks on a planet of Vega must wait until 2014 for the same experience.

Deneb must be altogether different. Since it is known to radiate with such intensity, it must lie far beyond its neighbors in order to appear no brighter than Altair. This star is so removed that its distance is not known precisely. The best estimates place it somewhere near 1500 light-years away, and thus we speculate that we are seeing it as it appeared in the fifth, sixth, or seventh century! Attila the Hun may have been ravaging the Roman Empire, or maybe King Arthur was convening the round table at Camelot, while farther east Belisarius scored another victory for the Byzantine Empire, or possibly Mohammed was fleeing Mecca on his famed Hegira, at the time of Deneb's appearance to our eyes tonight.

Yet even in A.D. 3500, when our descendants will see the Deneb of today, it will look just as it does now. Astrophysicists have pieced together a remarkably coherent picture of the lifetimes of stars, and the changes they undergo require much longer periods of time than have been mentioned here. But Deneb's constancy over time does not detract from the fact that we can still look up and see, in a sense, a little bit of the Middle Ages.

Clearly Deneb is a supergiant of a star to be seen easily over such a distance. Indeed, Deneb is so bright that it puts forth more radiant energy in a single night than does the Sun in an entire century! If Altair and Deneb were to change places, Deneb would appear almost as bright as the first-quarter Moon (when it is seen half illuminated). At 60,000 times the solar luminosity, it would be easily visible in the daytime and would cast shadows at night, whereas Altair could not be seen even through binoculars.

And now a word about the system of stellar magnitudes, by which astronomers measure the brightnesses of the stars. More than 2000 years old, the system is tried and true. It came about when the astronomer Hipparchus made a catalog of the stars visible from his home in Alexandria, the great port city of ancient and modern-day Egypt. Hipparchus, who lived in the second century before Christ, was not the first to undertake such a project, but his was a precise catalog, and it happens to be the oldest of those not lost to us.

Hipparchus included for each star not only its accurately measured position, but also an estimate of its luminosity. He divided the stars into five brightness classes called magnitudes and assigned the 20 brightest stars to the first magnitude. The next 50 or 60 brightest were of the second magnitude, and so on to the fifth magnitude, which comprised the stars that he could just barely see. His catalog contained just over 1000 stars, about half the number visible at any one time on the best of nights. Hipparchus's system remained in use until the nineteenth century, when observers realized that the eye, and the newly developed camera as well, records luminosity logarithmically. Astronomers used this fact to refine and extend Hipparchus's magnitudes. The refinement is based on the realization that a star appearing five magnitudes brighter than another looks to be 100 times as luminous. Using this relation as a precise ratio, they knew that each increasing magnitude represented an increase in luminosity of just about 2.5 times. Since telescopes allow us to see stars much fainter than the fifth magnitude, we need to extend the system to magnitudes 6, 7, 8, and beyond in order to include them. The system can also be made more precise by including tenths and even hundredths of a magnitude as needed. The result is a measure wherein the faintest star just visible to the naked eye on a clear, moonless night is of magnitude 6.0. Stars of magnitude 5.0

are about 2.5 times brighter then those of magnitude 6.0, and those of magnitude 4.0 are 2.5 times 2.5 (or about 6) times brighter than those of magnitude 6.0, and so on. If we reexamine the magnitudes of the brightest stars, we find out that a few stars as well as the Sun, Moon, and the five planets visible to the naked eye are so bright that the system must be extended to magnitude 0 and even into negative numbers.

Of the stars mentioned here, the brightest two, Arcturus and Vega, are actually of magnitude 0, not 1; their magnitudes are 0.0 to the nearest tenth. Altogether, there are 12 true "first"-magnitude stars; that is, 12 stars have magnitudes between +0.5 and +1.5. Spica, Altair, and Deneb are among these stars. Eight stars, including Vega and Arcturus, have magnitudes between +0.5 and −0.5 and can be said to have magnitude 0, and two stars are of magnitude −1. These are Sirius, visible only in the wintertime, and Canopus, located in the deep southern sky and thus not visible at all much farther north of the latitude of Atlanta or Los Angeles. Many of the planets are yet brighter. Jupiter shines between about −2 and −3, and Venus appears near −4. Mars varies greatly and can appear anywhere between −3 and +2. Saturn is fainter, being of either the first or the zero magnitude, and Mercury is always so close to the Sun that it is rarely seen, even when it is very bright.

Six of the seven stars of the Big Dipper are of the second magnitude, ranging from 1.8 to 2.4. Only the star joining the handle of the dipper to its bowl is fainter, being of the third magnitude. Polaris shines at magnitude 2.0 as does the brighter of the two bright stars at the other end of the Little Dipper. The fainter star is of the third magnitude. The other four completing the Little Dipper are all of about magnitude 4.5, hence their invisibility on any but the darkest of nights.

Three of the five stars forming the W of Cassiopeia are of the second magnitude and the other two are of the third, whereas only one of those belonging to her husband, Cepheus, shines at the second magnitude, and most of the others are only of the fourth magnitude.

On a clear, moonless night in the country, the fourth- and even the brighter fifth-magnitude stars are noticeable, but against the brilliant glare of city lights, few stars fainter than third magnitude are easily seen. Thus constellations formed of these fainter stars are not easily seen or recognized.

It was mentioned that about 20 stars are of the first magnitude or brighter; in modern terms this means that they are brighter than 1.5. The number of stars more luminous than 2.5, that is, the first- and second-magnitude stars considered together in the entire sky, is about 80, and the total brighter than 3.5 is about 300. When the fourth- and then fifth-magnitude stars are included, the totals rise to near 1000 and 3000, respectively. As each successive magnitude interval is added to the total, the number of stars triples.

The commonly given number of stars visible in an ideally dark, moonless, haze-free sky is 2500. This is a little less than half the total of 6000 in the entire sky that are brighter than the limiting threshold magnitude for the average human eye. The shortfall lies in the fact that at any time, half the stars are below the horizon, and near it some haze is always present obscuring the faintest stars. In the city, even with a clear horizon, the visible number is much smaller. How many fewer depends on the brightness of the sky due to light pollution. But a typical suburban night sky today allows only about 200 to 500 stars to be seen at one time in a clear, unobstructed, moonless sky, or about one-tenth to at most one-fifth the number visible in the same sky free of lights. It is primarily for this reason that old star guides are inadequate for the urban resident.

Chapter 5
The Summer Sky, Part 2

When I heard the learn'd astronomer;
When the proofs, the figures, were ranged in columns before me;
When I was shown the charts and the diagrams, to add,
Divide, and measure them;
When I, sitting, heard the astronomer, where he lectured
With much applause in the lecture room,
How soon, unaccountable, I became tired and sick;
Till, rising and gliding out, I wandered off by myself
In the mystical moist night air, and from time to time,
Looked up in perfect silence at the stars.

—Walt Whitman

The remaining summer constellations of interest include, from west to east and north to south, a crown, a harp, two birds, a dolphin, a scorpion, and a centaur.

The figure on page 42 shows the sky midway between the western horizon and the zenith. In it the constellations Boötes and Corona are seen, along with the Y-shaped figure formed from their brightest stars. The lustrous Arcturus lies at the foot of the Y. The arms are formed by the two next-brightest members of the kite-shaped Boötes and, to the upper left, a second-magnitude star lying near the kite but not part of it. Gemma is the brightest of a small semicircle of stars forming the constellation Corona Borealis, the northern crown. Another even fainter semicircle lies far to the south and is appropriately named Corona Australis,

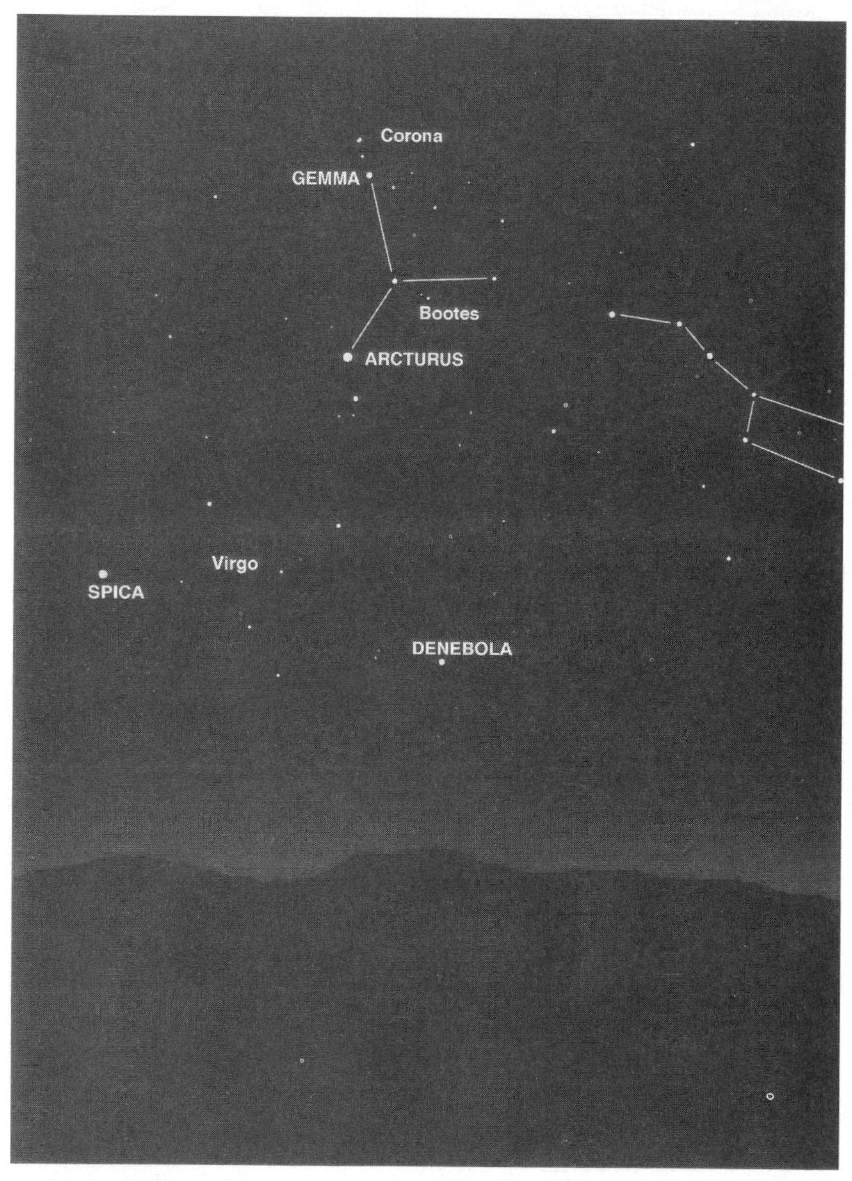

The midsummer view toward the west, showing the constellations Ursa Major, Boötes, and Corona Borealis and the bright stars Arcturus and Spica.

the southern crown. Our northern crown is the nicer of the two and is one example of a grouping of stars whose individual lackluster (with the exception of Gemma, all are of the fourth magnitude) would make them uninteresting if they were scattered over a large region of the heavens. The crown's compactness and attractive shape render it memorable, if not particularly lustrous, as a tiara should be.

Above and to the left, or east and south of Boötes and Corona are two of the least attractive and faintest constellations sprawling over immense portions of sky: Hercules and Ophiuchus. Who among us is not familiar with the legendary figure of Hercules, epitomizing magnificent proportions and feats of strength? His legendary prowess entitled him to a section of the celestial realm, and a large one at that. But the backyard to which he is relegated is outshone by almost all of his celestial neighbors. One neighbor who does not stand out is the equally large Ophiuchus, a star group pictured as a man holding a giant serpent. Neither man nor serpent plays a central role in classical mythology, although Ophiuchus was generally identified with Aesculapius, the god of healing, a mythical doctor or medicine man with whom serpents were commonly associated. From this setting, the snake led to the caduceus, the modern symbol of the medical profession. Aside from a fairly bright second-magnitude star making a triangle with Vega and Altair, the group is undistinguished. The serpent forms a separate constellation appropriately named Serpens, the only one in two separate pieces, much like the United States following the admission of Alaska, and Pakistan before the secession of Bangladesh.

It is time to assign the three bright stars of the summer triangle to their own respective constellations. Brilliant Vega dominates its small constellation as few other stars do. Lyra, the lyre or harp, consists of Vega and a small even parallelogram extending to the south of it. Deneb, on the other hand, is the first among near equals forming a large Roman style cross. The constellation is known as Cygnus, the swan. But the cross is familiar to many as the Northern Cross to distinguish it from the more renowned southern one. Deneb at the head of the cross lies at the tail of the swan flying southwestward. The bottom star of the cross is thus the head of the swan and lies just south of Lyra. Altair is immediately identifiable as the middle star of a small row of three. The three form the central part of an eagle known appropriately enough as Aquila. This aquiline creature flies northeast, just opposite in direction to Cygnus. Refer back to the figure on page 35 in Chapter 4 to see the major features of these three constellations.

Vega, or sometimes Wega, was named Allore or Alahore by the ancient Greeks, but like so many other stars, it is known by the Arabic name given to it some centuries later. It was the first star to be photographed, by the daguerreotype process, at the Harvard College Observatory on July 17, 1850. About 14,300 years ago, it was the pole star, the closest bright star to the North Celestial Pole. As discussed in Chapter 3, the motion called precession has carried the pole away from it, but 11,500 years from now, it will again be our pole star. It is by far the brightest of the bright stars that take turns serving as the pole star. Neither Vega nor its bright neighbors Altair and Deneb have the lively role in Greek or Roman mythology of Arcturus or the stars of Orion. In some Native American lore, Vega and Altair are lovers, forever separated by the Milky Way, which on dark nights away from city lights can be seen to pass between them.

Several small and very faint constellations surround these three, and any field guide to the stars covers them all. Only one among them stands out, and it does so for the same reason as does Corona; the proximity of the stars enhances their brightness—none is brighter than magnitude 3.5. Smaller than Lyra, Delphinus, the dolphin, resembles a small musical note. It can be found just above and to the left of Altair.

In the southern latitudes, the two brightest constellations in the summer sky are Scorpius and Sagittarius. Many readers will recognize these names because they are among the 12 constellations or "signs" used by astrologers in casting their horoscopes. These two constellations, along with the other ten, form the zodiac, the name for the great band girdling the celestial sphere against which the Sun, Moon, and planets are projected (see the figure on this page). The word *zodiac* comes from the same root word as do *zoo* and *zoology*, because all but one of its constellations represent living beings, including eight animals and three people (or four, if Gemini—the twins—is counted as two). This trend holds for the entire heavens, dominated as they are by stellar mammals, reptiles, and birds.

One of the most distinctive features of our solar system is its flatness. The planets orbit the Sun in nearly the same plane, and

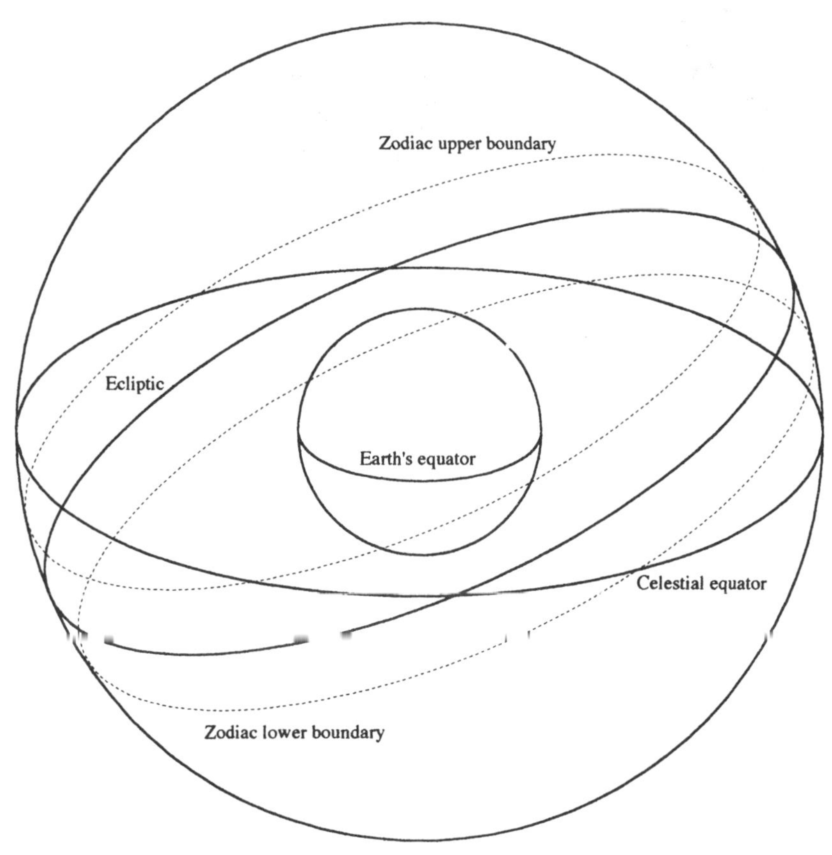

The relations between the celestial equator and the zodiac centered on the ecliptic.

as a consequence we see the Moon and the other planets confined to a narrow band centered on the ecliptic (the name given to the apparent path of the Sun through the sky as we orbit about it). The plane happens to be oriented in such a way that only certain constellations form the background. The lineup is a familiar one to readers of astrological columns in newspapers and begins traditionally with the point at which the Sun is located on March 21. There are problems with this lineup that the astrologers have ignored; these will be addressed in Chapter 26. The one worth noting here is that Ophiuchus occupies a section of the ecliptic and zodiac between Scorpius and Sagittarius (the astrologers' name for the former is Scorpio, but to astronomers it is Scorpius).

The twelve astrological signs, similar but not identical to their namesake constellations, begin with Aries, the ram, a small constellation of autumn. (Although the Sun passes through Aries in the spring, it is in the autumn when the ram is opposite the Sun in the sky that we can see it emerging in the east in the evening). In order follow Taurus, Gemini, Cancer, Leo, Virgo, Libra, Scorpius, Sagittarius, Capricornus, Aquarius, and finally Pisces. Six of the twelve are conspicuous or contain at least one bright star. These are Taurus, Gemini, Leo, Virgo, Scorpius, and Sagittarius. The other six are faint and will only be mentioned here.

Scorpius, the scorpion, is the brightest constellation around, with no less than six stars of second magnitude or brighter and nine others of the third magnitude. With all of the noble beasts that mankind could have placed in the sky, it may seem odd that this bright stellar assemblage should honor one of the least appealing. Scorpius is one of the few constellations that actually resemble the objects for which they were named. This group of stars looks like a scorpion. Below an upper portion of its head and body extends a row of stars curved just like a scorpion's tail, ending in a bright star where the stinger would be found. The brightest star lies in the body of the creature; it is the very red first-magnitude star Antares, whose name means "rival of Mars,"

the red planet. The stinger at the end of the tail is Shaula, which, at magnitude 1.6, just misses our first-magnitude list. It is the brighter of two stars just half a degree apart, sometimes called the cat's eyes. The tail hooks around under Shaula very low in the sky and can only be seen on the clearest of nights and not at all from latitudes north of Boston, Detroit, and Chicago.

Next door, just to the east is Sagittarius. This bright constellation looks like a teapot, slightly tilted to the right on which end is the spout, the handle being to the left. But in fact, it represents a centaur, a mythical creature whose upper half is a man and whose lower body, from forequarters back, is that of a horse, giving him six limbs in all. Sagittarius is called the archer because he is holding a bow and aiming an arrow directly toward the heart of the scorpion nearby. Although he extends just as far south as does Scorpius, no star brighter than magnitude 4 is located in this area; therefore all of the bright part, the teapot, is easily seen on a dark, clear night. These star groups can scarcely be seen from the British Isles. Antares is bright enough and at its highest is quite visible in the countryside from southern England at least, but Shaula in the tail is never visible there. A few other stars in the more northern parts of each of the two constellations are seen now and then, but that's all; Britain's night sights lie farther to the north. Both Scorpius and Sagittarius are shown in the figure on page 48.

No discussion of these two southerly members of the zodiac would be complete without a mention of the Milky Way, seen as a faintly luminous band coursing from Cassiopeia through Cygnus, passing between Aquila and Ophiuchus and on down to Sagittarius and Scorpius. In actuality, we know it to be a great spiral galaxy composed of a few hundred billion stars and many clouds of interstellar gas and dust. Our galaxy is at least 100,000 light-years in diameter and our solar system is displaced from its center by about one-third that amount. (Recall that light traveling at 186,000 miles per second covers just under six million million

The bright constellations Scorpius and Sagittarius are seen in midsummer along the southern horizon. Scorpius (right) resembles a scorpion and Sagittarius (left) a teapot.

The Milky Way arching overhead in midsummer. Courtesy of Yerkes Observatory.

miles in a year, defining a unit of distance.) At the center, or nucleus, of the galaxy lies the brightest part, containing most of the stars and total luminosity of the entire system. Extending outward from it is a thin disk of stars and gas all orbiting the nucleus just as do the planets around the Sun. All of the stars visible with the naked eye lie in our relatively small region of this disk, which is only a few hundred light-years in thickness. Each star revolves about the center in its own nearly circular orbit, slightly different from those of its neighbors. The Sun's orbit is typical; with the planets in tow, it loops around only once in about 220 million years. Since the age of the solar system and all of its members is believed to be about 4.6 billion years, it has completed only about 20 trips about the center of our galaxy since its formation.

The spiral galaxy known as Messier 74. Our Milky Way would appear like this if seen from far above its plane. Courtesy of Yerkes Observatory.

Located as we are within the disk of the Milky Way galaxy, we see the bulk of our stellar neighbors as a band circling the sky, which is in reality a great disk seen edge on. When Galileo first used a telescope in the year 1610 to look at the band, he discovered it to be composed of countless stars, most too faint to be individually visible to the naked eye.

The galactic center lies in the direction of the scorpion and the archer and far beyond the individual stars we can see. But it never gets very far above the southern horizon from our latitude, and is therefore never seen near full brightness above the low haze. Even on the best of nights, it looks no brighter than does the more northerly Milky Way in Cygnus and Aquila. But seen high in the sky from south of the Equator, our galactic center is so bright that on a moonless night it casts shadows! Our northern counterpart is a pale rejoinder to this southern spectacle. Still, it is the best we have. The band goes all around the sky and therefore passes through some of the winter constellations too. But the winter Milky Way is paler still and the sky is rarely clear and dark enough to see it at all.

Vega and Deneb each pass very near the zenith in the evenings of late summer, but this won't always be the case. Precession will carry both of them ever nearer the celestial pole. About 11,000 years from now Deneb will be the bright star near the pole, and 2000 years later Vega will have its turn, marking that point around which the sky appears to rotate. Vega was also near the pole 13,000 years ago and will mark it each 26,000 years, followed in order by Thuban, Polaris, Alderamin, and Deneb, among others.

* * *

One of summer's greatest celestial spectacles occurs every year around the tenth of August. This is the meteor shower known as the Perseids. Perseus is usually considered to be an autumn constellation lying right below Cassiopeia near the northeastern horizon in summertime. It is not a spectacular group but is still famous because it gives its name to this best-known stream of meteors.

To understand a meteor shower we need to think of comets in their orbits around the Sun. Unlike planets, comets' orbits are not almost circular but elongated so that the closest point of an orbit to the Sun, called the perihelion point, may be closer than any planet, and the farthest point, the aphelion point, may be among the farthest planets. Halley's comet provides a useful example. It moves in an orbit with a period of about 76 years, but at perihelion, when it is bright in our sky, it is close to Mercury's orbit and the aphelion lies close to Neptune's orbit, near the outer limits of the solar system.

Comets are known to consist largely of ice that vaporizes and forms a long tail as they approach the Sun and are affected by its heat. Embedded in the ice are tiny particles of stone and metal. After many passages near the Sun, a comet's ice is all melted away, but the particles remain and gradually spread out along the orbit. Whenever the Earth passes through the orbit, these bits of solid matter are seen as meteors streaking through our upper atmosphere and burning up from the heat of friction.

The Earth passes through such a situation every August 10 to 12. On those days meteors appear to stream toward us from a point in the sky, called the radiant, in the constellation of Perseus. Coming as they do at the height of the summer vacation season, they form not only one of the richest meteor showers of the year but also one of the most timely for viewing.

One does not have to see and identify Perseus in order to enjoy the Perseids. A clear sky free of trees and buildings as well as clouds and haze is all that is necessary to see about one meteor per minute by late evening. But it is a fact of life that a meteor shower—any meteor shower—presents more meteor events per hour as the night wears on. To understand the reason for this, we must consider the position of the Earth and our location on it.

The Earth orbits the Sun in the same direction as it rotates on its axis. Its orbital velocity is much the faster of the two motions, averaging about 20 miles per second, or some 70,000 miles per hour. This is 60 times the maximum rotational speed at the Equator. As the Earth plunges through a swarm of meteors, it is this orbital motion that determines the number of collisions. We are on the leading side or hemisphere between midnight and noon and at the lead point about sunrise. For the same reason, the trailing position is reached about sunset, at which time the only meteors we see are those that orbit the Sun faster than the Earth does and overtake it. But later on past midnight, some can come directly at us from the front. The number steadily increases throughout the night and reaches a maximum at dawn.

CHAPTER 6

The Stars of Autumn

Many a night I saw the Pleiades
rising thro' the mellow shade
glitter like a swarm of fireflies
tangled in a silver braid

—ALFRED LORD TENNYSON

Now, as the summer wears on, a new group of stars is slowly edging into our view by midevening. These are the stars in the eastern heavens, and they will dominate the zenith when the frosts of fall come and the nights begin to lengthen.

In the summer sky we found six stars of the first magnitude or brighter. Arcturus and Spica, actually stars of late spring, dominated the western heavens; Vega, Deneb, and Altair held forth near the zenith; and red Antares gleamed low in the south. The autumn sky is not so fortunate. Only one new star of such luminescence enters the stage. This is Fomalhaut, which is already making its appearance low in the southeast in the later evenings of August. Like Antares, Fomalhaut is far south and never gets very high in the sky. As for the rest of the autumnal lineup, we must be content with constellations formed of second- and third-magnitude and even fainter stars. This deficit will be made up with interest by the winter stars. But for now, the new star groups are more noteworthy for the legends they were intended to portray than for their luminosity.

An ancient Greek myth connects six constellations, of which Cassiopeia and Cepheus lie in the northern circumpolar region. The other four are all rising above the eastern horizon in the early autumn months. This myth runs as follows: Cassiopeia, a queen of Ethiopia and the wife of Cepheus, was very vain and boasted of her beauty and that of her daughter, the Princess Andromeda. Some of the gods, displeased by this display of vanity, expressed their displeasure by chaining Andromeda to a cliff by the sea, there to be at the mercy of a sea monster named Cetus. Enter Perseus, a certified hero for his slaying of Medusa, one of three gorgons with snakes for hair. Medusa had the antisocial ability to turn into stone anyone who gazed upon her. Perseus solves this problem by decapitating her while looking at her likeness reflected in his shield. Her blood drips into the ocean, and there gives rise to Pegasus, the winged horse. Pegasus does not serve as a mount for Perseus as some contend; that honor is saved for Bellerophon, who slew the monstrous Chimera while riding the winged steed.

Perseus arrives just in time to rescue Andromeda from Cetus by allowing the monster to see the head of Medusa or, in some of the accounts, by slaying the beast. Whether Cetus was slain or turned into stone (some contend into the Rock of Gibraltar, making Cetus a very large animal indeed), Andromeda was freed and her parents agreed to her match with Perseus. The lovers and Pegasus and even Cetus are all honored in the autumn sky, and together they dominate it. In early historic times Cassiopeia and Cepheus also appeared to rise and set in the autumn sky as seen from our latitude, but precessional motion has brought them closer to the pole and into the circumpolar realm.

* * *

The central feature and best-known asterism of this relatively undistinguished region is the great square of Pegasus. In midsummer at twilight it lies not far above the eastern horizon and it gets higher later in the summer. To find it, imagine a line extending from Polaris to the leading star of the W of Cassiopeia.

As one faces north, the dippers and the other stars appear to revolve counterclockwise about the pivot near Polaris. The leading star in the W is thus the topmost when it is to the right of the pole as it is in the summertime. Extend the line from Polaris through this star to twice the distance and come to another second-magnitude star. This marks the northeasternmost corner of the great square. Consider this star to be home plate; then the head of Pegasus extends out into center field beyond second base. The row of stars do rather resemble the head, neck, and forequarters of a horse seen upside down. The square is the central body, more or less, and the hindquarters and tail are just not there.

The star marking home plate is one of several stars scattered about the sky that is shared by two constellations, Pegasus and Andromeda. The young woman extends from this star in an arc in the direction away from Pegasus and consists of this and two other equally bright stars. Extend the arc to one more star; this third star is Marfak, the brightest star of Perseus. Perseus forms a constellation of no regular or memorable shape. But Marfak is very far north, so much so that it appears circumpolar from the most northerly large cities in North America and most of Europe as well.

Perseus contains one other star of the second magnitude most of the time. This is Algol, which really consists of two stars so close to each other that their period of revolution is less than three days (2.87 days, to be exact). Their orientation is such that each component star passes in front of the other and eclipses it. When the fainter companion passes by the brighter primary star, the brightness drops by more than a magnitude from 2.1 to 3.4; thus for a few hours every third day or night, Algol winks at us, only to brighten again later on. Once you are familiar with Algol's position in the sky, look for it each night. It is normally just fainter than Marfak, the other second-magnitude star of Perseus, but once in a while you'll see it appear much fainter than Marfak and several other stars nearby. Then you know that the brighter star is being eclipsed by the fainter of the close pair. Algol is the most conspicuous eclipsing binary star in the entire sky.

The other constellations of late summer and fall include four of the zodiac, all faint. These are Capricornus, the goat; Aquarius,

The autumn sky seen toward the south. The great square of Pegasus is near the center. Fomalhaut is the bright star near the southern horizon. Andromeda, Perseus, and Cassiopeia are also visible.

the water carrier; Pisces, the fishes; and Aries, the ram. Later on in the autumn, a large empty area toward the south marks Cetus, and in the deep south is the lonely first-magnitude star Fomalhaut. This star is part of a constellation known as Piscis Australis, the southern fish, not connected to Pisces of the zodiac, but no star other than Fomalhaut is visible except on the very clearest nights.

Fomalhaut plays a smaller role in classic mythology than any other first-magnitude star visible from the northern United States, Canada, and, on the clearest of nights, southern Britain. The reason for this is, again, precession. Two thousand years ago, this southerly star was much farther south than it is now, and only just visible in the skies of Athens and Rome. The great gyroscopic movement has lifted Fomalhaut, and all autumn stars, toward the north, as it does the spring stars to the south. Thus Antares, now just north of Fomalhaut and about equally conspicuous, was much farther north in Roman times, and figured more extensively in the old legends.

* * *

The autumn stars do not make much of an impression when viewed from the city, and none at all under bright lights. In the entire region only Fomalhaut and Marfak are brighter than Polaris at magnitude 2.0. The great square is the best-known asterism in the region and it is not overly conspicuous. We must wait until late in the fall, Thanksgiving at least, when the winter stars are well up, to see and experience the reason for this shortfall of bright stars. Before then, our bright summer friends begin to disappear. Spica and Antares disappear into the twilight by the end of September and Arcturus is gone a month later. The Summer Triangle remains visible almost into the Christmas season and shares the western sky with the stars of the Perseus legend.

As summer comes to an end, two new and striking objects appear in the east in the late evening. One of these two is Capella, a yellowish zero-magnitude star that is actually among the winter stars but is so far north and so bright that it dominates the northeastern horizon soon after it rises. By Labor Day it is visible at ten o'clock and a month later it is up by eight. It is part of the constellation of Auriga, which we'll see more of later. Capella is actually a triple star at least, with two bright stars orbiting each other, much as the planets orbit the Sun, and a more distant fainter star. Both of the bright stars are yellow, similar in color to the Sun, but they are giant stars, each more than 50 times as luminous. They are only half as far apart as are the Sun and the Earth, and they take but 104 days to complete an orbit about each other. They are about 42 light-years away from us and can be resolved as two stars only with sensitive resolution equipment.

The other early autumn spectacle is the Pleiades, a group of seven sisters of classical mythology. The Pleiades consist of a close grouping of stars that are not bright individually but are still conspicuous because they are so very close to each other. They are instantly recognizable and are hard to confuse with anything else in the sky; in fact, they may be the favorite star group of more people than any other. They rise with Capella and to the right, or south, of it, but being of the third and fourth magnitude, they are much fainter and must rise higher in the heavens to be seen through the haze.

This asterism constitutes a true physical group. Whereas most stars appear near each other in the sky because of a chance alignment, these lie actually in the same small volume of space and would appear huddled together seen from any direction. They form a star cluster known to have had a common origin about 75 million years ago and, due to mutual gravitation, have stayed together ever since. The cluster is a little over 450 light-years from us, and thus we see them as they were at the height of the Renaissance, when Copernicus, Columbus, and Leonardo da Vinci were creating our modern world. The Pleiades are part of the larger constellation Taurus, the bull, which is more properly discussed with the stars of winter.

The Pleiades present a very good test of eyesight. Many people see only the five or six of the brightest stars among them but some can spot seven, eight, or even nine stars in the group. Regardless of their number, they appear as a kind of tiny dipper sparkling in the eastern sky during the crisp evenings of autumn. As the season progresses, the Pleiades are seen ever higher until at Thanksgiving time midevening finds them high in the south. They represent a beautiful example of the fact that proximity between stars within a group is almost as important in their recognition as is their brightness. Because they are close, the seven sisters are as well known as Ursa Major, Cassiopeia, and Orion. Were they more widely separated, they would not be noteworthy.

Chapter 7
The Stars of Winter

Behold Orion rise, his arms outstretched encompass half the skies.

—Manilius

To those living in the warmer sunbelt, the winter sky may be just as familiar as the sky of summer. But Northerners probably spend much less time outdoors looking at the stars as the cold weather settles in. This is unfortunate because the onset of winter brings with it many of the brightest stars in the sky. The winter star groups, centered as they are on Orion, are matched in splendor only by those of the deep southern sky, invisible at northern latitudes.

For just this reason the Christmas holiday season is often portrayed as a period of crisp, clear weather with twinkling stars shining down on new-fallen snow. (Astronomers have made many observations with photoelectric photometers, instruments that measure sky brightness with very high accuracy. Their data refutes the presumed greater clarity of the sky in winter.) Since 1822 at least, when Clement Moore's famous poem "The Night before Christmas" first appeared, the image that comes to those of us who live in the north is of very clear skies. The denizens of the sunbelt may romanticize this image even though they have fled from the less attractive features of winter weather. The image seems incomplete without those brilliant stars shining in that clear black sky (with or without a full moon).

As late fall passes into winter, Capella in the northeast and the Pleiades in the east have risen to the point of visibility at sunset and dominance by late evening. They are the forerunners of a whole parade of first-magnitude stars culminating with Sirius, the brightest star in the night sky.

A very quick observation made on a clear dark night in the weeks before and during the Christmas holiday season suggests an alternate explanation for our impression of winter as the time of the clearest nights of the year. Some midevening in November or early evening in December, try making the observation for yourself. You need only a knowledge of the directions of the compass and a sky relatively free of trees, buildings, and bright lights, and—ideally—the Moon as well.

Imagine a line extending from the northern horizon through the zenith, and on to the southern horizon. This line, known as the celestial meridian, divides the sky into two equal halves, one to the east and the other to the west. The Pleiades should be immediately recognizable near this meridian, high in the sky toward the south. It should be immediately apparent to anyone facing south that the eastern half is spangled with bright stars, whereas the pallid western sky offers only a few stars and none of any brilliance. The stars to the west are the stars of autumn, and they will fade into the sunset as the winter wears on. The stars of winter are collectively much the brightest visible at our latitude.

We can get an idea of the imbalance by considering the 50 brightest stars in the sky. A list of these 50 includes stars all brighter than magnitude 2.0—the magnitude of Polaris, almost the faintest star on the list. The constellations of winter contain 17, or about one-third of the total. The autumn sky includes only 3, the lowest number for any season. (Five of the remaining stars are found in the constellations of spring and seven belong to those of summer. The north circumpolar region accounts for 4 more, and the rich southern sky contains the remaining 14 bright stars.)

Three features account for the predominance of bright stars in the wintertime. First, most stars in the Milky Way are found in the thin, central disk. From our position within the disk, we see stars projected along a band girdling the entire sky, but brighter in some parts than others. The brightest parts are those toward the center of the great system of our galaxy, located too far south to be seen in all its glory, even when seen from the southern states. The winter sky is the next brightest because it lies along one of the brighter parts of this band if not at its center. The stars of autumn are found in the direction more or less perpendicular to the disk, where stars thin out quickly with distance.

The constellation forming the center of the group of bright stars is the second feature. This is Orion, the hunter, and after the Big Dipper, the best known constellation of all. Most of its stars are very brilliant members of an association, a loose cluster of bright blue stars. These luminous aggregates form some of the most conspicuous features of our galaxy, and Orion, being the closest to us, is the brightest in the sky.

A chance arrangement forms the third feature. The two closest bright stars in the entire sky happen to be located right in this winter region. Their proximity assures that they will appear bright to us. Sirius is the more luminous; lying only nine light-years away, it is the brightest star in the night sky, exceeded in brilliance only by the Sun and Moon and the brightest planets. Sirius is known as the dog star because it lies in the constellation Canis Major, the larger of Orion's two dogs. It is easily seen to the southeast of Orion. To its north lies the star Procyon in Canis Minor only 11 light-years distant. These two stars are so close to us that they could have happened to lie in any direction. Most bright stars are much more distant; thus their locations are more likely to be near the Milky Way. Their chance location here just adds to the winter spectacle.

* * *

Look again at the northern sky. Perhaps you have been observing it, night after clear night, and noticing the Big Dipper

The midwinter stars seen toward the south as they appear in a dark sky away from outdoor lighting.

wheeling around under Polaris with Cassiopeia rising almost to the zenith by Halloween. Now in the cold sky of winter, Ursa Major is rising again and the queen is settling in the northwest. This orientation is shown in the figure on page 66. Compare this arrangement to the figure on page 27 in Chapter 3 that shows the northern sky at midsummer. Notice that each of the two familiar star groups has moved into the position occupied by the other back in the summertime. During the next summer the great balance will be restored again, with the dipper to the west of the pole and the great W to the east.

* * *

As we noted earlier, the Pleiades, or seven sisters, form a small part of Taurus. They are often pictured in mythological literature standing on the back of the bull. East of the group lies a reddish first-magnitude star known as Aldebaran. On a clear night it appears at one end of a group of stars forming a V about three or four degrees on a side. The other stars in the V are all physically associated and belong to a star cluster not unlike the Pleiades. These are the Hyades, which lie about 150 light-years away from us. The Pleiades are about three times as distant, and since the brightest stars of each group appear to be of the third and fourth magnitudes, the most luminous stars of the Pleiades must be much brighter than those forming the Hyades. Aldebaran is not physically associated with the Hyades cluster but happens to lie in the line of sight between the Hyades and us at about half their distance.

The Hyades are an older, looser, and fainter cluster than the Pleiades. If the two clusters were to change places in the sky, the Pleiades would contain at least four stars of the first magnitude, more than are found in any present constellation. The Hyades, in contrast, would be inconspicuous; their brightest stars could not be seen without binoculars. Aldebaran, at a distance of 60 light-years, would not be affected by this transition.

Among the writers of horror and fantasy literature, the Hyades have long been portrayed as places of evil. They are populated by some of the sinister Old Ones in the mythos of H. P.

The midwinter stars seen toward the north.

Lovecraft and his followers in that genre. Lovecraft's fictional author, the mad Arab Abdul Alhazred, in his book the *Necronomicon*, refers to the dark stars near Aldebaran and the Hyades, to which the Old Ones fled when cast out for their nefarious practices here on Earth. Perhaps the first appearance of the Hyades in the evening sky near Halloween gave rise to their association with the eldritch, the squamous, and the blasphemous.

To the north and east of Taurus, Capella shines brightly yellow. Capella, another of the zero-magnitude stars like Arcturus and Vega, lies at the zenith in the constellation Auriga, portrayed as a man holding a goat. Although a bright affair, it lacks a memorable shape. At best we can call it a pentagon, though not a regular one.

Below Taurus and rising in the east is Orion, the giant, known also as the hunter and the central one, and perhaps after the Big Dipper the most widely known star-figure of all. Orion, sometimes Arion or Urion in times past (and sometimes blithely written as O'Ryan by the Irish), is missing a proper head and feet, but his torso is clearly delineated. Four stars account for two shoulders and two knees and a tight row of three other stars forms a belt at his waist.

Five of this group of seven are among the brightest fifty stars, the most belonging to any constellation. Orion possesses not one but two of the ten brightest of all; the zero-magnitude star Betelgeuse marks his left shoulder, and the slightly brighter Rigel marks the right knee. Taken together the two make perhaps the most noticeable color contrast among all stars. Rigel is distinctly blue, and Betelgeuse is one of the reddest bright stars in the sky. The color of a star is set by its surface temperature; red is the color of the coolest stars, and as we proceed from them to the hottest, we encounter the colors orange, yellow, white, and finally blue.

Orion is a star factory. Look just below the belt to see his sword, composed of two fainter stars and a fuzzy spot in between. This patch is not a star at all; it is a nebula, the brightest in the sky. Most of the stars of Orion are young—millions, not billions, of years old—and this nebula at Orion's center is still churning out stars today. The gas and dust shine by reflection and by reradiation of energy from the hot young stars cooking inside.

Betelgeuse may be the most famous of star names. The name is of Arabic derivation and purportedly means "armpit of the central one," a reference to its location in Orion, and his central location with respect to his bright neighbors in the sky. Some stars (Sirius, Capella) have Latin names, but most (including Vega, Altair, Deneb, Fomalhaut, and Aldebaran) have names that were assigned by the people of the Middle East as they kept astronomy alive and flourishing during the European Dark Ages.

Arabic names sound unfamiliar to most of us and allow for inconsistencies in their spelling and pronunciation, Betelgeuse being a prime example. "Beetle juice" has become the most amusing and fashionable pronunciation, but conscientious astronomers and other scholars have tried others including BETT-lejuice, BettleGURZ, beTELgus, BetaiGOITze, BAYtlegus, and others. By whatever name, this star is a veritable jumbo as stars go; if placed at the location of the Sun, the orbits of Mercury, Venus, Earth, and Mars—the whole terrestrial realm—would all lie within its limb. With only 10 to 20 times the Sun's mass, this dull red blob is a near vacuum, at least near its surface. By any measure, "the armpit of the central one" is a fascinating star.

Rigel is the rival of Betelgeuse within Orion, but it has little of the intriguing nature of the red star. Twice as distant as Betelgeuse, this blue supergiant is, like Deneb, one of the very brightest stars in this part of the Milky Way galaxy. Outdone by Betelgeuse in interest and nearby Sirius in sheer luster, Rigel may not always receive the attention due the seventh brightest star of all, and one that is so intrinsically luminous that it could be seen almost halfway across our galaxy.

Winter is, by its nature, a time of long nights. The stars are visible well before dinnertime and by the time the late news appears on television the sky has rotated through more than one quarter of a full revolution. Which, then, are the stars and constellations that are most appropriately assigned to winter? The division by season is inevitably arbitrary, but a split of the major portion of the celestial sphere that is neither circumpolar nor perpetually hidden from us can be made such that in midevening the stars surrounding the meridian as seen in the middle of the respective season (approximately the first days of February, May, August, and November) constitute the ones for that season. At twilight in early winter, the autumn stars still dominate the zenith, and near winter's end the stars of spring have all but taken over. Even one summer asterism remains prominent until Christmas. Cygnus, the swan, with its bright star Deneb standing atop its northern cross, is still visible low in the northwestern sky. Those who have remained familiar with the aspects of the Summer Triangle region will have followed the northern cross all the way from the zenith to the horizon where it is still just visible. Above the cross lies the autumnal host of the Perseus legend, followed by Orion and his retinue.

Whenever Orion is high, the origin of his epithet as "the central one" becomes obvious. An aura of brilliant stars surrounds him on all sides. The conspicuous winter constellations form a large, roughly circular region with Orion at the center. We are already familiar with Taurus above him and to his right, with Aldebaran as its lucida, or brightest star. To the left, or east, of Orion lie the two brilliant stars Sirius and Procyon. With Betelgeuse they form an almost perfect equilateral triangle, sometimes called the Winter Triangle (see the figure on page 64 in this chapter). Procyon, a zero-magnitude star, is white and located due east of Orion's famous shoulder. Procyon lies in Canis Minor, the small dog, a constellation with but one other visible star nearby. But it is to Sirius that all eyes will be drawn.

There is no mistaking Sirius for anything else. Over four times as bright as Betelgeuse, Rigel, or Procyon, it twinkles at magnitude −1.4. Sirius is the brightest star in the night sky and the brightest blue object of any kind in the sky. At a distance of only nine light-years, it is almost a next-door neighbor. It is almost 30 times as luminous as the Sun, bright but much less bright than the far more distant stars of Orion. Four other stars join Sirius to form a kind of stick dog with forepaws to the right of Sirius and three bright stars forming the hindquarters and tail below. The brightest of these three southerly stars, Adhara, is itself just barely bright enough to be a first-magnitude star. Adhara is pretty far south; it lies farther south than Antares and just a bit farther north than Fomalhaut. Precession is moving Fomalhaut northward and Adhara in the opposite direction at such a rate that Adhara will replace Fomalhaut as the most southerly first-magnitude star visible in Great Britain and the northern United States in the year 2096, about a century from now. And Sirius, itself, is moving south as well. Precessional motion will carry it beyond the southern horizon of London about A.D. 7800, and it will disappear from the skies of New York some 1800 years later. After that time, the three brightest stars of the sky and the only three of negative magnitude will all be invisible from the cities of Europe and northern North America for many thousands of years.

Sirius enters our lore in a number of ways. It cannot be seen between mid-May and mid-August, for at those times the Sun is north of it, rising first and setting last. Then in late August it can be seen rising just ahead of the Sun in the dawn. These hottest weeks of summer were known to the ancients as the dog days since Sirius and the Sun were close together in the sky at that time and their combined radiation was thought to make for the increased heat of that time of year. In ancient Egypt, the date when Sirius could first be seen rising in the dawn coincided with the annual flooding of the Nile River, a very important event for that desert land, even though Egypt at the time of the building of the pyramids (about 2700 B.C.) was less arid than it is today. The Sahara Desert was then more of a grassy plain, and about three times as much water flowed in the Nile and flooded its banks then as now. The sacred priests, responsible for such things, looked for Sirius in the dawn sky just before sunrise. Morning after morning they failed to spot it, until the Sun moved just far enough to the east for the star to be seen glimmering in the dawn. This date is called the heliacal rising date for Sirius; the heliacal setting date is the last evening in which it can be seen at dusk. For the latitude of New York, the dates of last setting and first rising fall within a day of May 14 and August 17, respectively; from London, they will occur a few days earlier in May and a few later in August, making the period of Sirius's absence longer. In ancient Egypt they fell earlier and closer together because Egypt lies farther south and precession has moved Sirius eastward with respect to the Sun since that time. The scene is shown schematically in three steps in the figure on pages 72 and 73.

* * *

Sirius twinkles more than any other object in the sky. All twinkling, sparkling, and scintillation is a result of the turbulence in our atmosphere. To understand the process of twinkling, we must be aware of the intrinsic difference between stars that twinkle and planets that do not, unless they are near the horizon. Stars shine by their own light and are many thousands of times brighter per unit area than planets, much as the Sun is about half a million times as bright as the full Moon, although they appear the same size in the sky.

Planets rarely twinkle, because they are not self-luminous; thus, they appear to have much larger angular diameters than stars. Their light is integrated or smoothed over the breadth of their disks. As one light beam gets bent one way, another gets bent in the opposite direction canceling the effect. The stars, none of which appear as disks in even the largest telescopes, are too small for the smoothing to take place, and the starlight shimmers this way and that as a result. The brighter the star, the more twinkling can be seen, and Sirius is the brightest.

The angular difference between planets and stars is great—the brighter planets can be seen as disks with binoculars, but even the most brilliant among the stars appear as points when viewed through our largest telescopes. Pluto shows by far the smallest angular diameter of any planet, yet it appears larger than Betelgeuse, which shows about the largest disk of any star.

If we imagine the celestial sphere shrunk down to the size of the Earth and mapped onto it, we can gain an immediate appreciation for this difference in angular size. On the Earth's surface, one degree would measure about 70 miles, so the Sun and Moon would each cover a circular region 35 miles in diameter. Jupiter,

The heliacal rising of Sirius at three different stages. As the sun moves eastward each day, Sirius rises further ahead of it, and finally rises sufficiently earlier to be seen in the dawn sky. Courtesy of Griffith Observatory.

Continued.

somewhat brighter than Sirius, would vary in size, averaging about three-quarters of a mile across. Contrast this with the size of Sirius in our model. It would be about the size of a dinner plate, less than a foot across!

Sirius can be seen almost on the horizon on a clear night. It never ascends above about 30 degrees in altitude as seen from the latitude of New York, or above 20 degrees from London, and is thus never seen high in the sky. Whenever it is particularly low near the horizon, the twinkling and scintillation become more pronounced, flashing through every color of the rainbow. In general it will appear red, rather than its normal blue, for the same reason that the Sun and the Moon appear redder when they are seen near the horizon. This is one possible explanation for a number of ancient and medieval observers reporting a red color for Sirius. We have solid astrophysical evidence that Sirius would once have appeared red like Betelgeuse, but the time scale would have to be much longer than the time covered by recorded history.

Another explanation for this enigma has often been offered. Sirius has a white dwarf companion star, a faint, dense planet-sized star that is known to have once been a red giant, much as Betelgeuse is today. A white dwarf is known to be the end product of a star that has gone through its intricate evolutionary process to a point where its outer layers have sloughed off into space and its core has collapsed under its own gravitation to form a small, very dense substar. Sometime in the past, this dwarf was the much brighter star and the unresolved pair would have been dominated by its red light. After a cataclysmic event the outer layers would have been blown off and the small white dwarf star is all that would remain. Again, the time required for this to happen is much longer than the historical record. It is infinitely more likely that the red color reported by these observers was due to proximity to the horizon. Still, these reports are a bit surprising since any experienced observer would be aware that Sirius is blue whenever it is seen well above the horizon. Mysteries such as this one persist throughout science, and we can only hope in the future to make sense of them.

CHAPTER 8

The Stars of Winter, Part 2 

Other constellations complete the ring around Orion, the central one. Canis Minor has been mentioned and rather briefly dealt with, as befits a group of only two stars—Procyon and a third-magnitude star lying about five degrees above and to the right of it. Unlike its larger companion, Canis Minor can scarcely be seen to represent any animal. Still, no culture could ignore a star as bright as Procyon, and so it became a dog.

Between Orion and his two dogs and filling most of that large triangle formed by Sirius, Procyon, and Betelgeuse is a sprawling constellation known as Monoceros, the unicorn. Here is a good example of a region ignored by early peoples, and not until the time of post-Renaissance Europe, after the invention of the telescope, did this gap get filled by the likes of Monoceros. If this celestial unicorn were elsewhere in the sky, it might receive some attention. But between the stars of Orion and his two dogs, with four of the ten, and ten of the fifty brightest stars in the whole sky, it has little chance for attention. On a clear, moonless night it is easy to spot several fourth-magnitude stars within the triangle; however, they are all but lost amidst the splendor of their brilliant neighbors.

Unlike Monoceros, Lepus is a smaller and somewhat brighter constellation that is worth attention. Lepus, the rabbit, is just below, or to the south of, Orion and just to the right, or west, of Canis Major. A not insignificant compact group of third- and fourth-magnitude stars, the rabbit passes through the sky directly ahead of the larger dog, which is sometimes portrayed as chasing it.

Directly above Canis Minor and east of Taurus lies a group that is hard to ignore. Two stars of nearly equal brightness stand together at the head of this group; they are Castor and Pollux, the twins, dominating and giving the name of Gemini to the whole constellation. At 34 light-years distance, Pollux is the brighter and redder star, and it is actually somewhat closer to us than is blue Castor by almost 20 light-years. Together the two form the closest bright pair in the northern sky and deserve their designation as twin stars. Gemini, like Taurus, is one of the twelve constellations that girdle the sky and form the zodiacal background against which the Sun, Moon, and planets are seen.

The glory of the winter stars is viewed by some as a giant ring (this can be seen in the figure on page 64 in Chapter 7). With Betelgeuse in the center, the other bright stars form a near circle about it. Capella is near the Zenith, and looking clockwise from it, we see in turn Aldebaran, Rigel, Sirius, Procyon, and finally Pollux and Castor. The spectacle is enhanced if Mars, Jupiter, or Saturn happens to be shining in their midst.

If you should happen to see the winter sky on a moonless night and at a location from which the sky is very dark, you may notice a faint band passing through Auriga and proceeding to the south and east through the Winter Triangle and Monoceros. This is the winter Milky Way, forming a paler but much smoother stripe than the portion of the Milky Way visible in the summertime.

The Milky Way of winter does not line up with the brightest stars of Orion and the other bright winter asterisms any more than does its summer counterpart with the stars of that season. It seems that the brightest stars and constellations form a band of their own tilted nearly 15 degrees with respect to the fainter unresolved stars that make up the Milky Way. This band formed by the bright stars is called Gould's Belt, after Benjamin A. Gould,

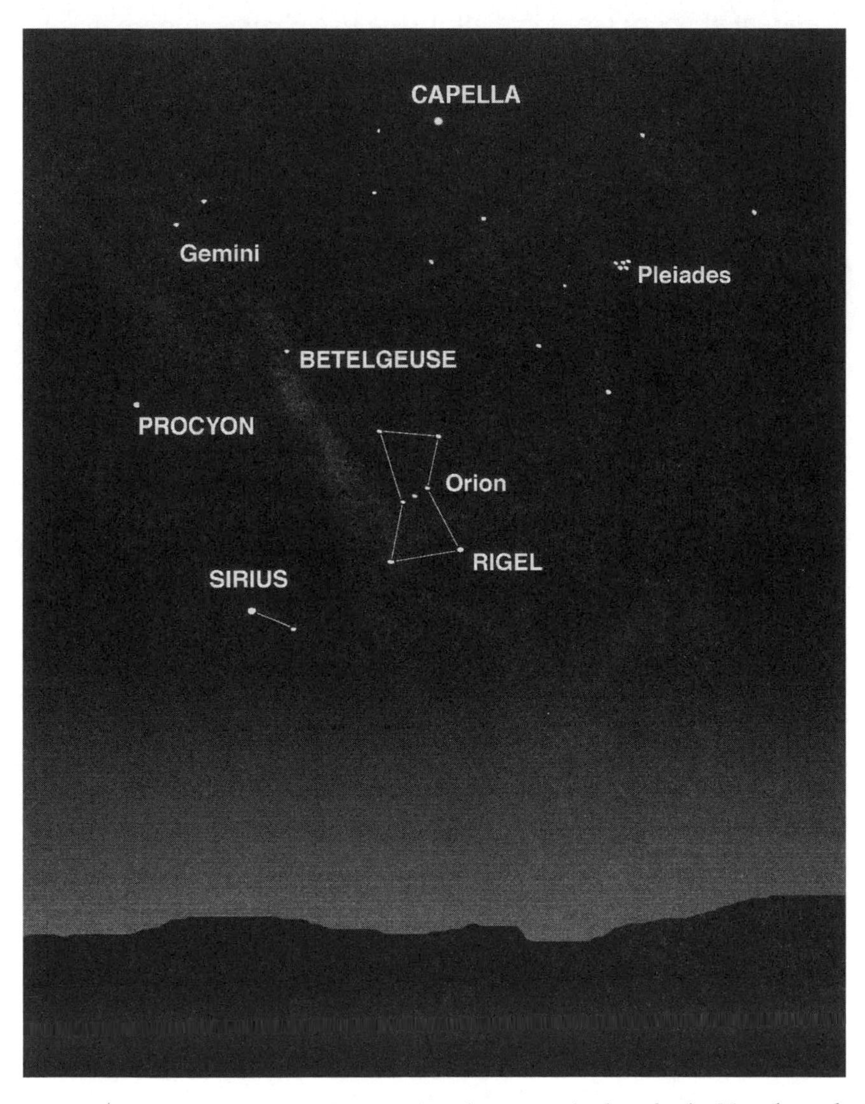

The midwinter stars seen toward the south as they appear in the suburbs. Note that only the bright stars are visible, and none are much closer to the horizon than Sirius. Compare with the figure on page 64 in Chapter 7.

an American astronomer who studied the misalignment about a century ago. Astronomers have since learned that much of the apparent effect of Gould's Belt is due to a chance alignment of a few extended physical groups of stars called associations. They are not to be confused with clusters such as the Pleiades and Hyades, despite the coeval formation of their stellar members. Clusters occupy a much smaller volume than do associations, and continue as recognizable groups far longer, before their stellar members dissipate into the galactic background. Associations are often found near nebulae and other interstellar gas and dust. Their stars are very bright and very young and have not long ago condensed out of the interstellar material. Orion is much the closest and most spectacular of these associations; as already mentioned, within it is a giant star cloud churning out new stars even today. The center of the group is in the fuzzy patch of the nebula, midway along the sword of Orion hanging down from his belt. Most of the bright stars forming the constellation are associated with it. The whole lot lie at a distance of about 1500 light-years from us and are celestial infants, probably only a few million years old.

The winter sky requires great clarity for many of its finer sights. In the city and suburbs, the sky rarely appears as crisp and clear as is depicted in the figures on pages 64 and 66 in Chapter 7. Much more commonly it resembles the figure on page 77, where many stars are faint or invisible altogether. Notice the absence of the three tail stars of Canis Major and all of Lepus. The visible stars are faint in a sky that appears washed out. This city–country difference is of concern to modern skywatchers, and will receive full attention in Chapter 11.

CHAPTER 9
The Stars of Spring

Shortly after Canis Major is fully visible in the southeast, Leo puts in an appearance in the east. Thus he is well up in the evening sky by midwinter. Leo is a constellation of two parts, each a well-recognized asterism. The head and front quarters of the lion resemble a sickle. When it is high in the sky, the sickle is upright and the handle extends downward to the first-magnitude star Regulus. The sickle rises in the east on its side and is followed about an hour later by three stars forming a right triangle. This latter group represents the hindquarters and tail of the beast. The brightest of the three is Denebola, a second-magnitude star at the tail. Some of the stars in the sickle and the triangle are not very bright, but the two together make up one of the best-known constellations in the sky (see the figure on page 80).

North and east of Leo, rising by midevening in February, is our old friend Arcturus. Arcturus is the first of the seasonal stars (those that rise and set) to return. We made our first acquaintance with this bright orange star in midsummer. Then it was descending into the west, not long after sunset. It slipped into the twilight sky in October, and before Thanksgiving it was gone. Now it is back, the first noncircumpolar star to reenter the celestial stage. It is a brilliant reminder that the Sun circles the sky each year, and different parts of the celestial sphere show up in their turn when the Sun has moved elsewhere. Boötes is now seen to lie nearly parallel to the horizon, its kite-shaped form stretching to the

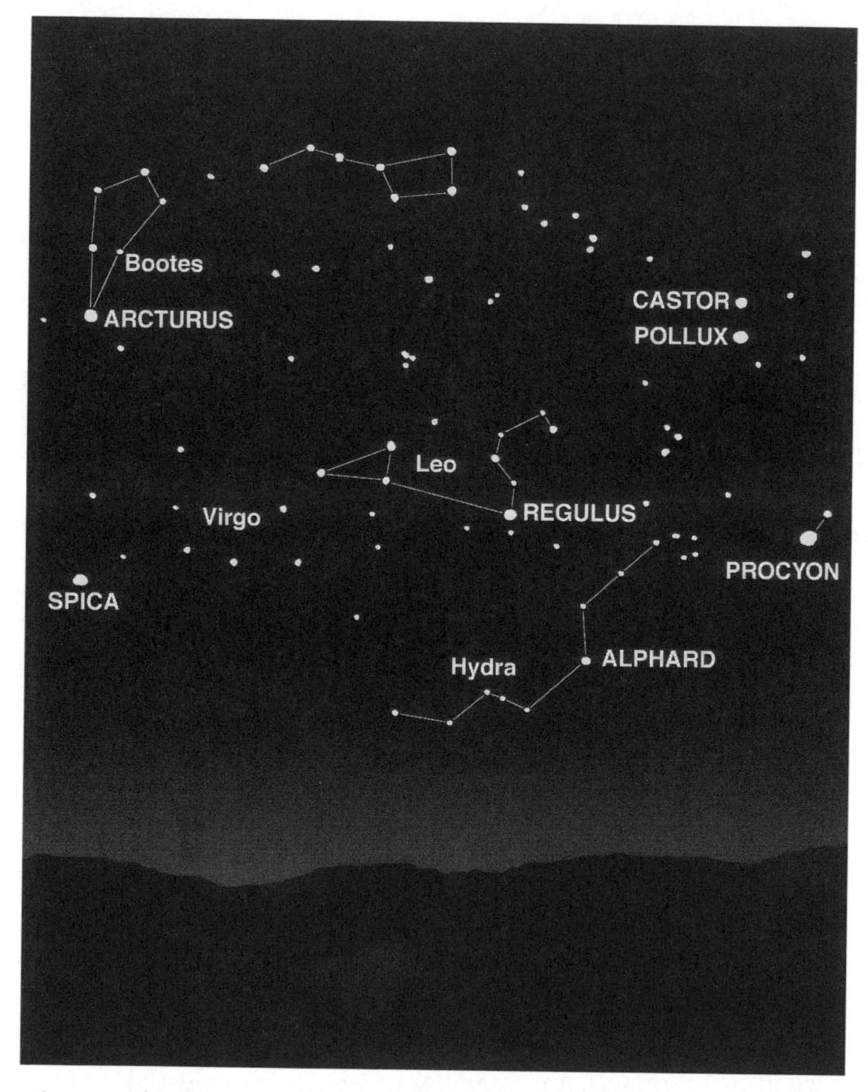

The stars of early spring seen toward the south. Leo is in the center of the figure. Arcturus and Spica are to the left, and the twins and Procyon are to the right.

north, or left, of Arcturus. Corona Borealis rises just after it, tagging along as it did in the summertime.

Arcturus dominates the springtime, the season of rebirth, as does no other star in no other season. Of zero magnitude, it is hardly brighter than the summer star Vega or the winter stars Capella, Rigel, Betelgeuse, and Procyon, and it is but a quarter as luminous as Sirius. But unlike the gaggle of stars brightening the winter sky, Arcturus is alone, particularly after those bright stars have settled into the twilight and before Vega ascends high in the early summer sky. Only if Jupiter or Mars is nearby is Arcturus upstaged in the eastern sky of March and April.

Arcturus and Denebola in the tail of the lion form a nearly perfect equilateral triangle with Spica, another old friend from the summer. A little larger and less luminous than the Winter Triangle, this trio deserves its name of the Spring Triangle. Spica is the lucida, or brightest star, of Virgo, a large and sprawling constellation of the zodiac with very few other bright stars and no recognizable shape. Four reasonably bright stars lie to the right, or west, of Spica. Their outline suggests the sail of a sailboat, about six degrees in length and height. This is actually the constellation Corvus, the crow. Corvus reminds us again of the attractiveness of proximity; its four stars are close to each other and are all equally bright, being between magnitudes 2.5 and 3.0. Spica and the three next brightest stars in Virgo are just as bright as these, but as a group, they are too widely scattered to attract as much attention.

When Leo is near the meridian, look to the west, where Procyon can still be seen. Just above and to the left of that star is a small ring of stars depicting the head of Hydra, a giant snake. Hydra winds on to include Alphard, the lone bright second-magnitude star below Regulus in the sickle of Leo, on well south of Denebola, on past and just south of Corvus, on below Spica, and on and on for almost one-third of the way around the entire sky. Not all parts of this behemoth are illuminated with stars, but it is easily the longest of all constellations and is also the largest in area. With the exception of the circlet of stars forming the head, and the lonely Alphard, Hydra has nothing for the naked eye. The front part of Hydra is below and to the right of

Leo in the figure on page 80. The figure on page 83 shows the sky about a month later for the same time in the evening. Here, Hydra's tail is just visible below Spica and to the left of Corvus.

One theory to account for Hydra's great length and relative emptiness has it that when the original 48 constellations were named, presumably about 4000 years ago, the stars forming this giant snake lay along the celestial equator of the time, and thus marked it. This idea is plausible, if not proven correct.

Aside from Leo, Boötes, and the crown, almost none of the constellations of spring have bright or very memorable outlines, and Hydra and Virgo are no exceptions. The brilliant winter sky is fading fast now, with only Capella and the twins left to linger in the late spring twilight. The skies of spring look drab and faint, as do their autumnal counterparts. Both are humbled by the dazzle of Orion and his dogs to the west, and to a lesser extent by the emerging Summer Triangle to the east. I previously noted this feature and the reasons for it (Chapters 7 and 8); it is an effect of the orientation of the plane of the Milky Way and of Gould's Belt.

Two other constellations are worth mention. Both are most easily found by noting voids in the sky with few visible stars. The first is just to the right, or west, of Regulus and the sickle of Leo, between them and Gemini, the twins. Here lies Cancer, the crab, and, along with Pisces, the faintest of the 12 zodiacal constellations and astrological signs. Below and to the east of Boötes and Corona Borealis sprawls Hercules. He was also with us last summer as much as he ever is—faint indeed as seen from the city. Hercules, like his fellow hero Perseus, is associated with other star groups lying nearby. Among these are Leo, Cancer, and Hydra. According to Greek myth, Hercules was a demigod, having as he did a god, Zeus (or Jupiter), for one parent and a mortal for the other. Jupiter's goddess wife, Hera (Juno), was jealous of the whole affair and required the strong man to perform 12 very arduous tasks or labors for Eurystheus, a king of Mycenae in her favor. The first called for Hercules to vanquish a ferocious lion that had been having its own way in the valley of Nemea. He was then to take the skin of this beast to Eurystheus. The second task was to slay the terrible Hydra, a monster who had the annoying ability to grow two new heads to replace any one that Hercules

The stars of late spring seen toward the south. Boötes and the crown are near the upper left and Corvus is low near the southern horizon.

cut off. In time he managed to slay both creatures as well as a crab sent by Juno to bite him. Now Leo, Cancer the crab, and Hydra are all honored in the sky not too far from Hercules.

As spring wears on, we will see the bright Scorpius with its lucida, the red first-magnitude Antares, rising in the southeastern sky. Antares is a red supergiant similar to Betelgeuse. This "rival of Mars" lies close to the orbit of that planet. Whenever Mars lies close by, the two form the brightest red pair in the sky.

Just above and ahead of Antares are two stars of equal brightness pretty much by themselves. They form about all there is to see of Libra, the scales. This is the only constellation of the 12 forming the zodiac that represents an inanimate object. At one time near the dawn of recorded history, it was considered part of Scorpius, forming that creature's front claws. But for the last 30 centuries or more, it has been its own constellation. Perhaps its greatest claim to fame lies in the names of its two bright stars, Zubeneschamali and Zubenelgenubi. These are tongue-twisting Arabic names for the northern and the southern claws of the scorpion.

With Libra we have covered all 12 of the signs of the zodiac. The 12 astrological signs are similar to but not identical with their namesake constellations. The signs are considered by most practicing astrologers as fixed with respect to the colures, the two equinoxes and the two solstices. Thus the astrological lineup will always begin with Aries and proceed through Taurus, Gemini, Cancer, and the rest, ending with Pisces.

This is the arrangement as it was aligned at the height of classical times and for the Roman Empire, as well as for the beginning of the Christian era. But in the intervening two thousand years, precessional motion has carried the colures backward through the width of about one sign or one-twelfth of the entire ecliptic. Today the vernal equinox point is located in western Pisces, and the daily horoscopes should begin there. But during Roman times, this first point lay in Aries, about to enter the sign of the fish, and the early Christians may have taken their symbol from the coincidence. Now the equinox is about to leave

Pisces and cross over into Aquarius, thus initiating the "Age of Aquarius."

Opposite Aries in the sky, Libra was the home of the autumnal equinox two millennia ago, and still is according to most astrologers. But having moved with the rest, the autumnal equinox has since backed into Virgo.

CHAPTER 10

The Southern Stars

What do northerners envision when they think about nights of tropical splendor? Whether from the beaches of Florida and the Caribbean islands or from the deck of an ocean liner cruising through southern seas, our image is of clear, balmy nights under skies brilliant with stars and perhaps the Moon languidly passing behind a small cloud or two. Is this vision based on reality, or is it a figment of the imagination of those whose business it is to lure us from the cold northern cities into the warmth and charm of the south? For the most part, the answer is that this image is real.

As we have seen, those of us who reside in the northeastern or midwestern United States and Canada are used to an appearance of the sky that is summer-oriented. Summer skies are pale with few bright stars, and they are often hazy from the presence of dust and water vapor. Winter skies contain many more bright stars, but clear winter nights are cold and few of us spend much time outdoors looking up at them. Occasionally, around the Christmas season, shoppers, skiers, and carolers may pause on a chilly night to view Orion and his two dogs. But most of our celestial gazing is done in the warmer months. It is then that we are most likely to renew our acquaintance with the Big Dipper and the other familiar star patterns.

Christopher Columbus and other ancient mariners knew that the altitude of Polaris and its well-known neighbors varies with one's latitude. As they sailed northward or southward over the curved rim of the globe, these star groups appeared higher or lower, respectively, in the night sky. Today's traveler to the south will share this same experience. The fact that this experience occurs at any longitude is itself a proof of the sphericity of the Earth. If the Earth were flat, the same stars would be seen by everyone.

From New York City lying at latitude 41 degrees north, the North Celestial Pole, with Polaris lying nearby, is located 41 degrees above the northern horizon. Any constellation found within 41 degrees of the pole is circumpolar; that is, it does not rise or set but is visible all night, every night. From London, more than ten degrees farther north, the pole is seen more than ten degrees higher in the sky, and the circumpolar area is larger. Similarly, all of the sky within 41 degrees of the South Celestial Pole is never seen from New York, no matter how bright it may be.

These are not coincidences. The altitudes of the poles in the sky are directly related to, and indeed fixed by, the latitude of the observer. Travel east or west merely changes the time at which an object is seen, but travel north or south changes the appearance of the sky itself.

Imagine traveling steadily northward from New York toward our North Pole. As you proceed, you will see Polaris rise higher and higher in the sky, and stars in the southern sky will sink ever lower until they disappear below the southern horizon. At the pole itself, the circumpolar area encompasses half of the entire sky, and the other half remains perpetually hidden from view. Sirius, Rigel, and Spica are lost to view, and Betelgeuse, Procyon, and Altair—only a little farther north—are low but still visible on clear nights. Orion is split in two, with stars above the belt circling the horizon while those below remain invisible. Half of the sky is seen all of the time.

Now suppose you were sailing south. Night after night the pole and Polaris would sink toward the northern horizon, and the stars to the south would rise. These southern stars would be followed by others that inhabit the sky remaining unseen from New York. If it was springtime, the Southern Cross would rise from the waves on an ocean voyage toward the tropics.

At the equator, no stars are circumpolar and none remain invisible for more than 12 hours. Each celestial pole lies on the horizon, and even Polaris passes a degree above and below the north point in a tiny circle every day. Every star is above the horizon for 12 hours and below it for 12 hours. All of the sky can be seen half of the time.

Contrary to popular belief, the educated elect of our civilizations have known for about 24 centuries that the Earth is round. The spherical Earth concept is due in part to Aristotle, who lived in the fourth century B.C. He even offered a proof for this shape, one that is still accepted as a proof today. Eclipses of the Moon are actually passages of it through the shadow of the much larger Earth. Such events happen every few years, and in every case the edge of the Earth's shadow as it crosses the Moon shows the shape of an arc segment of a circle. From his observations of this feature, Aristotle correctly reasoned that the Earth is a sphere since that is the only shape that casts a round shadow from every direction.

Eratosthenes (ca. 276–195 B.C.) went further. He lived in Alexandria a century after Aristotle and was a predecessor of Hipparchus, whose star magnitude system gave rise to our own. Eratosthenes not only knew the Earth to be round, he also obtained a nearly correct value for its size. On a trip from Alexandria to Syene in southern Egypt (now called Aswan, site of the great Aswan Dam) Eratosthenes noticed that the Sun illuminated the bottom of a well in late June. This meant that the Sun in Syene reached the zenith at noon on or about June 21, the date of the summer solstice. It did not do so when seen from Alexandria; the length of the shadow of an obelisk there at its shortest showed that the Sun at its highest was still seven degrees south of the zenith. Eratosthenes knew that Syene lay 5000 stadia (a stadium was a unit of length equivalent to about a tenth of our mile) to the south of Alexandria; thus 7/360 of the Earth's circumference equaled 5000 stadia in length. He obtained a value for the circumference that is anywhere from almost exactly the correct size to about three-quarters as large as we know it to be. The uncertainty here results from our inability to retrieve through translation the number of stadia in a mile. This was a great achievement by any measure, and the size of the globe has been known to mariners since that time.

At no time in the past and in no other place have more people been able to take advantage of the experience of a major shift of latitude as is now possible in eastern North America. Once familiar with the sky at home, tourists heading to Florida and points south find it an enriching experience to discover new stars and constellations and to view the familiar ones in new orientations. Westerners would experience the same effect on a trip to Hawaii or central Mexico, and English travellers would do so on a tour to the Mediterranean Sea and points south.

A trip from Boston to Washington, or from Chicago to Saint Louis, or even from San Francisco to New York for that matter, is not sufficient for this because in each case the change in latitude is only three degrees.

The Miami–Fort Lauderdale area lies at 26 degrees, and Tampa and Orlando at 28 degrees north latitude. Both are 12 to 15 degrees south of the urban regions of the east and Midwest, and anyone familiar with the bright stars at home can immediately appreciate the consequences of such a large difference. Farther south, Puerto Rico and the Virgin Islands, Jamaica and the Cayman Islands, Mexico City and Acapulco all lie at latitudes 17 to 19 degrees north, and the difference from the sky at home is even more noticeable.

From Florida southward, the Big Dipper and Cassiopeia are not circumpolar, but rise and set as do most of the rest of the stars we see. From anywhere south of the Equator, even Polaris is invisible. But there are compensations that offset the lessened visibility of these familiar stellar friends. As the northern stars lower, those in the south appear to rise, and new ones become visible. Surely the most famous among them is the Southern Cross. It adorns the flags of Australia, New Zealand, Papua New Guinea, and Western Samoa and is to the Southern Hemisphere what the Big Dipper is to the north. Its constellation name is Crux, and unlike the relationship between the Big Dipper and Ursa Major, the asterism and the constellation are one and the same. It is a bright affair; the two brightest stars are of the first magnitude, and the third is not much fainter. Only the fourth star

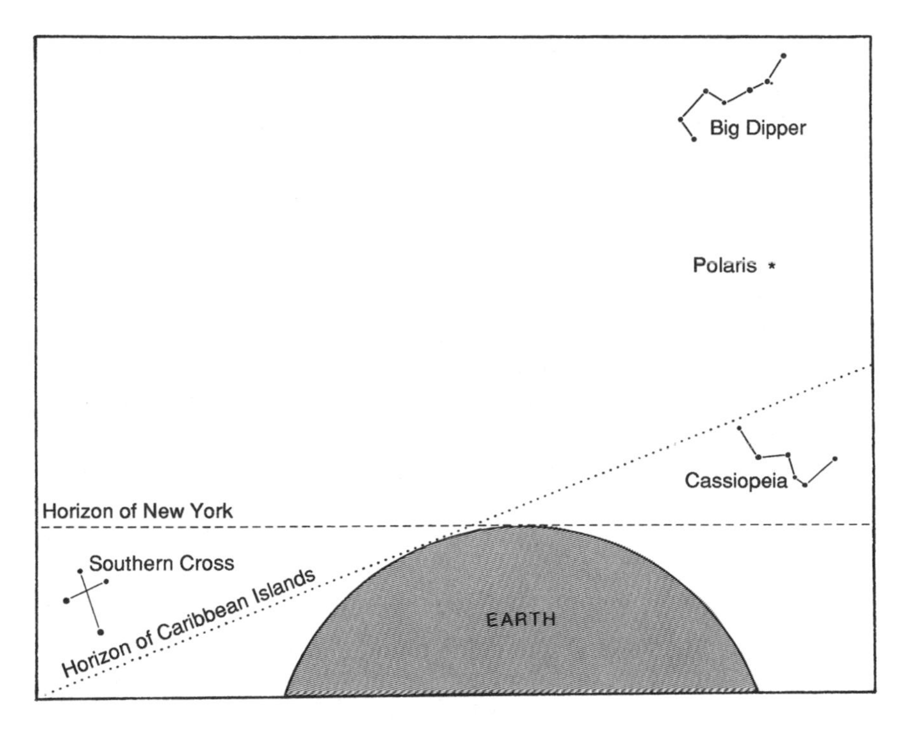

The horizon forms a tangent line that touches earth at the point where the observer stands. The figure shows the horizon at New York and in the Caribbean region.

spoils a perfect cross, being not merely fainter, but also somewhat out of position. From southern Florida, the top star of the cross can be seen just above the horizon, but from the latitude of Puerto Rico, Jamaica, Mexico City, and Honolulu, it is easily visible higher in the sky of late winter and early spring. Not far to the left, or east, of the cross, one can spot two very bright stars. These are the two brightest stars in the large constellation of Centaurus, the centaur, the mythical half-man half-horse of Greek legend, and they are known most commonly as Alpha and Beta Centauri. Alpha, the brighter, has the distinction of being the closest star to our solar system. It is actually a triple star, but the two brighter stars are seen as one without a telescope, and the third is much too faint to be seen without a telescope of some size. Only four light-years from us, these stars form the first destination of almost every fictional star trek that leaves the solar system for adventure or settlement beyond. At a magnitude of −0.3 for the combined light of its component stars, Alpha Centauri appears as the third-brightest star in the night sky.

Earlier in the winter in these southern climes, Orion and Sirius will be visible much higher in the sky. A subtle change comes over Sirius whenever it is viewed at or near the zenith. As noted earlier, it never ascends above about 30 degrees when seen from New York, but from Miami it can be seen more than halfway to the zenith. On a calm night Sirius does not twinkle very much. Up north we cannot see any blue object that bright and that high in the heavens with the comparatively steady light of Sirius seen from the south. This is one more difference between tropical and temperate skies.

Well below the brilliant blue Sirius (but nicely above the southern horizon) shines its only stellar rival, Canopus. This luminous white star, with Sirius, dominates the entire sky unless the Moon or a bright planet is visible. Canopus is so bright (at magnitude −0.7) that it can even be seen faintly just above the southern horizon from the latitude of Atlanta and Los Angeles. But from Florida and farther south it can be seen in its full glory.

This great star is not brilliant just because it is close to us, as is Sirius, only nine light-years off. Rather, Canopus is a true supergiant of a star, like Deneb, Betelgeuse, and Rigel. Until this decade, we didn't really know its distance; it was placed at only 100 light-years distant by some, and six times as far by others. Situated as it is in the deep southern sky, it has not been accessible for observation by the great observatories, which until recently have all been located in northern regions. But now, thanks to a new space satellite, we know its distance with high precision to be just over 300 light-years away from us, with an uncertainty of less then ten percent of that figure.

Canopus is the brightest star in a huge and brilliant stellar group representing Argo Navis, the ship in which Jason and the Argonauts sailed in search of the Golden Fleece, according to Greek legend. Like the Standard Oil Company, this vast constellation was too big, and early in this century, it too was broken up into several parts. Now divided into four modern constellations,

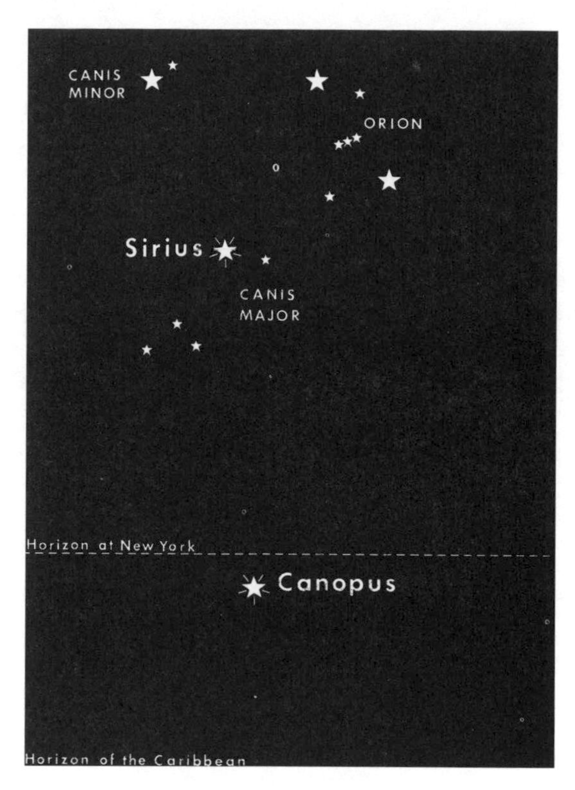

The stars of midwinter as seen above the horizon at New York, and from Florida and the Caribbean Islands showing the bright star Canopus directly below Sirius. Compare with the figure on page 64 in Chapter 7.

the giant ship lies along the Milky Way between Orion and Sirius to the north and the Southern Cross to the southeast. No other part of the sky can begin to match the number of bright stars contained in the region stretching southward from Orion and passing through Canis Major, the ship, the cross, and the centaur. These constellations lie along one of the great spiral arms of our Milky Way galaxy within which most bright stars are found. Only from the tropics and the Southern Hemisphere can they be seen together in full.

One other bright star found in these southerly reaches can be seen from southern Florida just after dark at Christmastime. Its name is Achernar, meaning the end of the river. The river in this case is Eridanus, one of the longest constellations. It begins right next to Rigel and winds down to end at Achernar with no other conspicuous stars anywhere along the way. In classical times, Achernar was the most southerly of all bright stars, and the only one visible neither within the entire Roman Empire nor within the empire of Alexander the Great. It is not surprising that it played no part in the great classical mythologies of those times. Achernar is moving north on the great precessional arc and is today just visible in Alexandria and Jerusalem. In another 20 centuries, it will have nudged its way into the skies of Rome and New York as well.

A trip to the tropics also changes the way we see the Moon. In the midnorthern latitudes, the Sun and the Moon both rise and set moving from east to west, due of course to the rotation of the Earth. From lands near the equator, the Sun and Moon are seen to

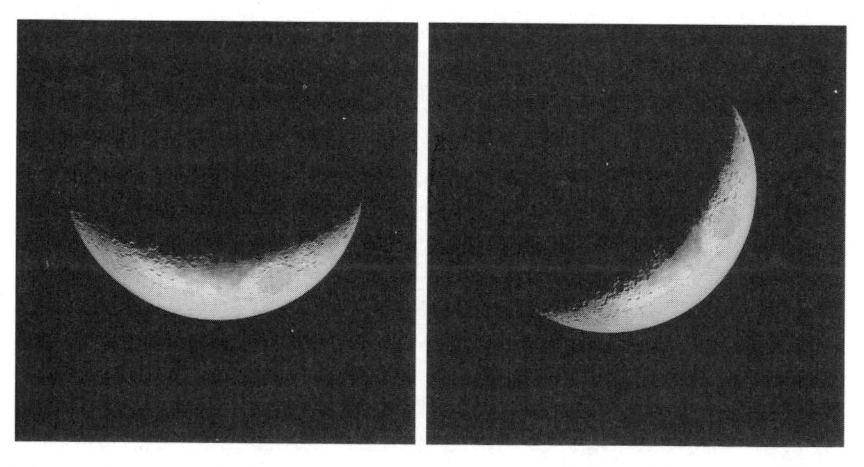

The waxing crescent Moon as seen from the tropics (left) and from midnorthern latitudes (right). Courtesy of Yerkes Observatory.

rise high in the sky, and can even pass through the zenith at times, after which they descend toward the western horizon in a path almost directly perpendicular to it. This motion gives rise to two effects that seem unnatural to denizens of the north but that form the heart and soul of the nighttime tropical experience.

If we consider the Moon in its crescent phase and recall that it is illuminated by the Sun, it is easy to see that the points, or horns, of the crescent will always point away from the Sun. As the crescent Moon moves across the sky following the Sun, its horns point nearly straight upward and it will appear in the evening after sunset as a small shining boat gently sinking toward the horizon in the western sky.

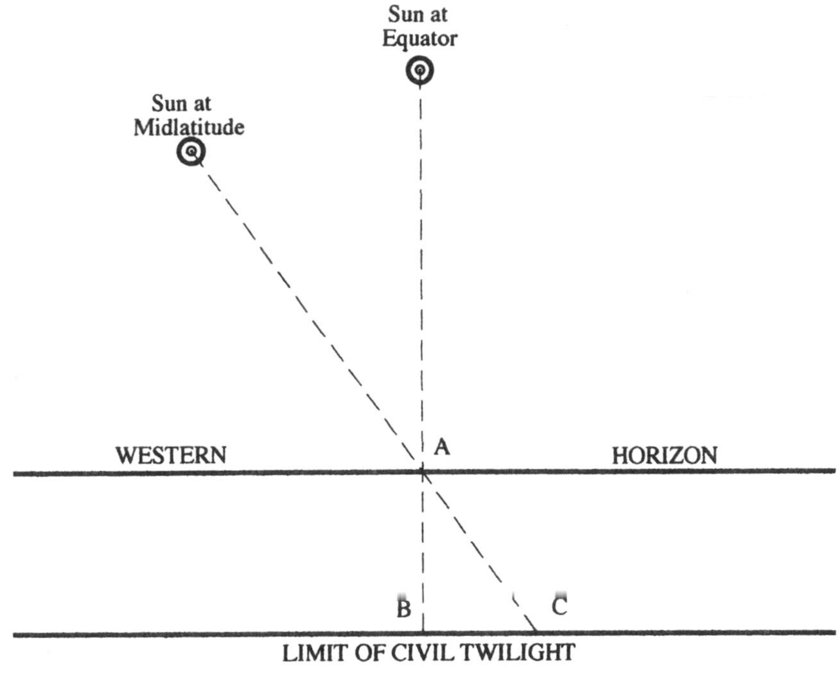

The path of the Sun near sunset as seen from the equator and from midlatitudes. It covers the space between the horizon and the limit of civil twilight below it in a shorter time when seen from the equator (line AB) than from midlatitudes (line AC). Hence the twilight sky gets dark more quickly at the equator.

From north temperate latitudes, we see the crescent Moon as it appears on the right in the figure on page 94, and never as it appears on the left. First-time travelers to the tropics notice the unusual orientation at once.

The perpendicular motion of the Sun has the effect of shortening the time between sunset and total darkness. The duration of twilight is simply a function of the apparent motion of the Sun, specifically the angle at which it approaches the horizon. When descending straight downward, as it appears to do in the tropics, it gets to the level below the horizon at which darkness is apparent in a shorter time interval than if moving toward the horizon at an angle (see the figure on page 95). Night comes in a rush in the tropics as a result.

Taken together, all this helps to define the experience of a tropical evening. Darkness falls moments after sunset, and if the time of month is right a silvery crescent Moon sails in a celestial sea like a small boat just above the spot where the Sun has disappeared. In the vividly clear velvety darkness, hundreds of bright stars suddenly appear, many more than Northerners are accustomed to seeing in their skies at home. The swarm of bright stars makes the tropical nights appear clearer than they actually are. Sun, Moon, and stars taken together redeem the promise and lure of a clear night in the tropics.

Part II

The Wanderers— Sun, Moon, Planets

CHAPTER 11

The Brightness of the Night Sky

There were lights, thousands of them, but of no brightness: these were thousands of tiny flecks affecting the darkness not at all; they were gaslights, most of them, white at this distance and almost steady; but there was candlelight, too, and I supposed, kerosene. No colors, no neon, nothing to read, just a vast blackness pricked with lights . . .

So writes Jack Finney in *Time and Again*, his glorious and loving portrait of the New York City of 1882 as seen through the eyes of a late-twentieth-century time traveler. Even at that time, New York was a great metropolis of almost two million people, most living in Manhattan.

For centuries, cities have illuminated their streets and walkways at night. In ancient Rome torches were placed along some of the lanes around the Forum. From that time until the last century, cities have used the same methods to provide light for outdoor evening activity. In the nineteenth century, gaslight predominated, as the largest urban areas, led by London, began to grow into the millions for the first time. The old lamplighter, familiar in song and legend, made his rounds as he lighted each streetlamp in turn. Then, just over a century ago, Thomas Edison and his coworkers invented an incandescent bulb in what may have been the first of all research laboratories. He illuminated a street in New York City with a row of arc lights, and the modern era began.

It is not easy for us now to imagine the comparative darkness in even the largest cities before the advent of modern streetlights and advertising signs. Each change of streetlighting—from gas to incandescent bulbs to the present mercury and sodium vapor lamps—produced a wholesale rise in nighttime brilliance. Even the movies that recount the gloomy London streets of late Victorian times, haunted by the likes of Sherlock Holmes and Jack the Ripper, portray the lighting levels of the time much too brightly for us to appreciate the full impact of Finney's observation.

Still, cities of the last century were noticeably brighter to their inhabitants than the countryside, as Vincent van Gogh noted in a letter to his brother, Theo from *The Complete Letters of Vincent Van Gogh, Volume 2*:

> One night I went for a walk by the sea along the empty shore. It was not gay, but neither was it sad—it was—beautiful. The deep blue sky was flecked with clouds of a blue deeper than the fundamental blue of intense cobalt, and others of a clearer blue, like the blue whiteness of the Milky Way. In the blue depth the stars were sparkling, greenish, yellow, white, rose, brighter, flashing more like jewels, than they do at home—even in Paris.

For half a century, streetlights and private and advertising lights were lit with the incandescent bulb, almost unchanged since Edison's time. Then, in the years following the Second World War, more economical sources of light became available. Indoors, fluorescent lights competed with the tungsten bulb. Outdoors, the mercury vapor lamp quickly came to cast its bluish glow above highways, byways, and even farmyards all over America and much of Europe. These lights cost less to operate than their predecessors for the same amount of illumination but otherwise gained us little in pedestrian or vehicular safety. Tubing filled with rarefied glowing neon and other gases provided illumination for advertising purposes, but this form of light has proved too fragile for more widespread use.

In subsequent years most mercury vapor lights were replaced in turn by the still less costly sodium vapor lamps that now glow across the landscape and seem to be in universal use. They emit an orange-pink light, a blend of light over the yellow, orange, and red portions of the spectrum, unlike the incandescent white lights composed of all of the colors. A minority of outdoor lamps are of the bluish metal halide type, most commonly reserved for situations where color rendition is critical.

As new lighting replaced old, three factors led to our present artificially brilliant skies and wasteful ways. First, in the drive toward energy-efficient light fixtures, those responsible for light installation and maintenance replaced existing bulbs with brighter ones of the same wattage, rather than ones that maintained the same light level with a lower power consumption. Second, lighting fixtures sent light out in all directions, including upward where it is almost never needed and serves only to illuminate the universe. Finally, the suburbs' constant expansion into the countryside dramatically increased the individual security lights found mostly around suburban and rural homes; this trend has brought about a light pollution problem of huge proportions.

Only in the last decade has the dubious maxim "more light is better" been discarded by law enforcement officials and utility companies. Studies reveal no connection between streetlight levels and crime rates, and if excessive outdoor lighting leads only to a false sense of security, we would be better served by more proven safety measures. Whatever further study may reveal, the overriding issue is not the worthiness of lighting as such (which is clearly necessary at some level), but rather its efficiency.

Who among us has not suffered the temporary blinding effects of glare from the setting sun when we drive in its direction? Glare is the result of light entering the visual field that is greater than the level of luminance to which the eyes are adapted, causing annoyance, discomfort, or loss in visual performance and visibility. Poorly designed streetlights can produce glare as well, creating potentially perilous conditions for pedestrians and traffic. An assailant approaching from the direction of a bright light may not be seen in time for his intended victim to take evasive action. Motorists blinded by direct glare from streetlights shining into their eyes cannot always detect a pedestrian or another vehicle approaching from the side, especially under misty or rainy conditions. Far more safe and efficient are downward-shining lamps with their light shielded from above and from the side. The figure on this page shows two walkway lamps, one of which is shielded. The illumination shining downward onto the ground is the same for both, but glare and upward lighting make the one on the right much more wasteful of electricity. The newer fixtures provide more uniform and less spotty illumination of a walkway or highway and can achieve the same light levels with a considerable reduction in the wattage used by the earlier light fixtures (see the figure on page 103).

The financial gains of conversion from wasteful to efficient bulbs and housings are well documented. The city of San Diego, for example, has realized a saving of over three million dollars annually, a cost reduction that continues in perpetuity. The recovery in the costs of the retrofitting of streetlight fixtures is accom-

Two walkway lights; the light on the left is shielded, thus eliminating upward light and glare. The light to the right shines down no more brightly but shows much wasteful glare.

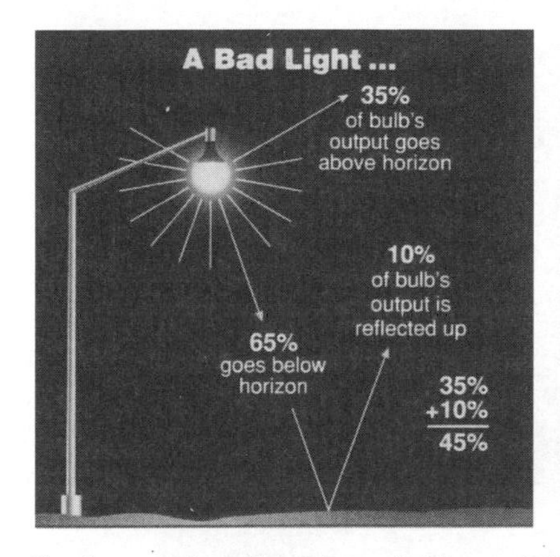

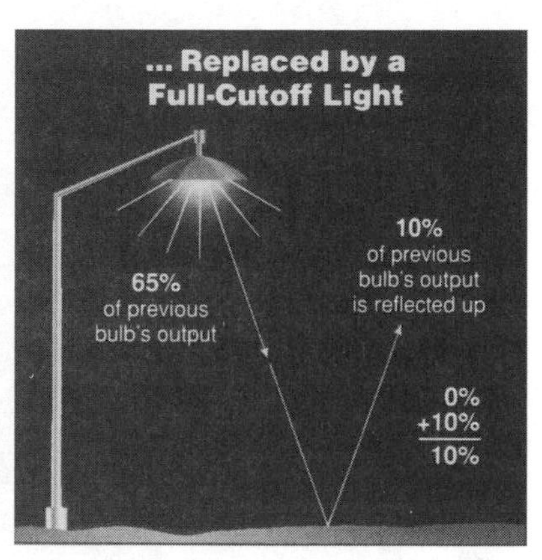

Skyglow can be reduced by switching to full-cutoff fixtures. With reduced wattage, the downward illumination remains the same, but shielding reduces upward-shining light by about three-quarters. Copyright Sky & Telescope, *1996.*

plished in about three years, after which time the savings are free and clear.

It is natural that astronomers were the first to be alarmed by the growth of outdoor lighting—they have the most sensitive instrumentation to detect it and to be inconvenienced by it. Astronomy with large telescopes requires very dark skies. Most telescopes are located on mountaintops in the southwestern United States and in Hawaii; consequently, the nearby communities in these regions were among the first to bring outdoor lighting under control. Our present bright skies deter every aspect of astronomy, from stargazing for pleasure to education and research.

Outdoor lighting has also become an ecological problem that is currently receiving attention from environmental groups. Ecological damage due to light pollution is less extensively verified than astronomical damage; however, it undoubtedly exists at some level. Some species of migratory birds depend on nighttime navigation for their seasonal travels and appear disoriented by the blaze from below. The circadian rhythms of some plants can suffer under 24 hours of continuous bright light. Bright nocturnal skies along the East, West, and gulf coasts disrupt natural cycles

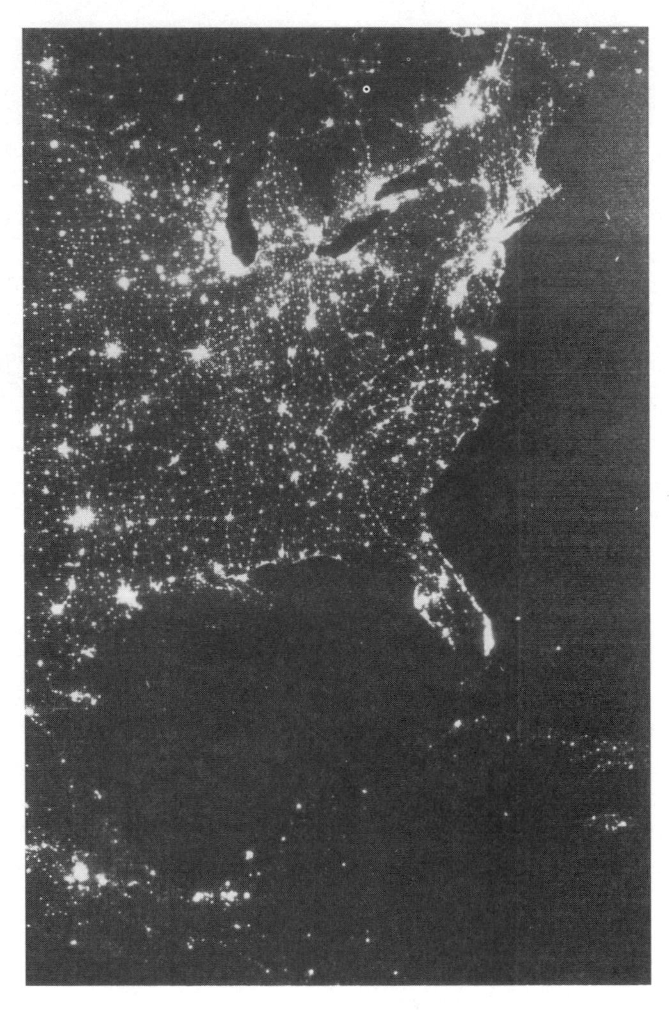

The Earth at night showing bright lights in eastern North American cities and towns. Courtesy of NASA.

there as well. The sky above the coastline is now brighter than that above the sea, whereas under natural conditions, the situation would be reversed due to bioluminescence present in the oceans. According to the Center for Environmental Education, the brightness differential provides a false direction for hatching infant sea turtles on the beach—and now often leads them landward to their deaths.

Above all else looms the financial waste incurred by current lighting levels. The present cost of light pollution is not yet widely known even among environmentalists, but estimates of its scale have been made recently by the International Dark-Sky Association (IDA) organized in Tucson in 1988 to find and publicize solutions to the problem of light pollution and to bring it under control. The conservative estimate of the cost of only the portion of outdoor lighting shining above the horizontal plane—some one-third the total nighttime illumination—is well over one billion dollars annually in the United States alone. Since the waste is in the form of electricity usage, it translates directly into the consumption of coal and oil, with the attendant addition to global warming through the emission of carbon dioxide and other greenhouse gases into the atmosphere. Light pollution has taken its place alongside more familiar afflictions, such as air and water pollution and the mismanagement of toxic waste, as a major despoiler of our environment, one that furthers our dependence on foreign oil. And it deprives over 90 percent of all Americans the right to behold the night sky in its full glory.

Within the last decade, several promising events signal a possible future slowdown in the surge of brightness. Many utility companies have shifted their positions on new and brighter lights from advocacy to neutrality. The Illuminating Engineering Society of North America and others have encouraged lighting manufacturers to produce fully shielded, energy-saving, reflective light fixtures. Outdoor lighting needs have become more carefully defined. Illuminating engineers are becoming as concerned with the overall cost of lighting as they are with personal safety, and traffic consultants are becoming aware that direct glare is not conducive to safe driving. Lighting engineers and ecologists far outnumber astronomers among the IDA membership, now that all interests can be mutually satisfied by lighting control and regulation. Nor can the IDA claim all of the credit for the growing number of statutes limiting the growth of light pollution; the laws strictly limiting streetlights and private lights along the gulf coast of Florida were passed because of efforts by environmentalists without astronomical input.

On a larger scale, the magnitude of the light produced by this wanton nighttime fecundity of human society is of a level visible from other worlds. By day, the Earth is strangely devoid of signs of human habitation, even when seen from the space shuttle only a few hundred miles overhead. Great cities appear as featureless dusty patches against a more fertile green background, and once in a while a dam or levee appears as a telltale straight line.

It has been said often enough that the Great Wall of China could be seen with the naked eye from the Moon, the only human artifact to be so favored. It is indeed long enough; it would stretch across half the face of the Moon if it had been built there. But what about its width? Only as wide as a two-lane highway, it is far too narrow to be resolved from nearby space, much less from the distance of the Moon. Many parts of the American interstate highway system would be more visible, if length alone were the determining factor. But they aren't visible either.

Surprisingly, evidence of human activity makes little visible impact on the appearance from space of the sunlit side of our world. But at night communities of all sizes are ablaze with light. Slash-and-burn fires destroying the tropical rain forests of the Amazon River basin, fishing fleets in the Sea of Japan, oil field fires in Siberia and the near east—these shine forth as does no other dark sunless place in the solar system. Even New York City, with its giant skyscrapers and its Great White Way, appears dominated by the baleful orange glare of millions of streetlights, as does every other sizable urban agglomeration on the planet. From the Moon the naked eye could only too easily make out the northeastern corridor from Boston to Washington as an orange patch of sodium vapor light. Cities of the third world appear almost as bright. Only the mighty atmospheres of Jupiter and Saturn and the seething surface of the Sun show changes of greater magnitude than the ever-brighter night side of the Earth.

New York City and environs from space. Long Island is seen to the lower right and the coastal cities of Connecticut as far as Bridgeport are seen in the upper right. Courtesy of NASA.

Light is measured in a bewildering variety of units, and we can define the brightness of the night sky in terms of any of them. Furthermore, the natural illumination of the night sky is variable and is due to at least four different sources. These are described in Chapter 2; they include the faint airglow of the upper atmosphere, the reflected sunlight from dust and debris between the planets, the light from faint unresolved stars, and the very faint light from interstellar nebular material. Fortunately, their combined average is well known, and it serves our purpose here to define the average night sky at one unit of luminance. By doing so, the average suburban neighborhood is found to be about 5 to 10 times this value, whereas city centers may range as high as 25 to 50 times the natural level. The numerical value is given for the zenith or point directly overhead, because that is normally the darkest region of the sky. Sky brightness increases, often in a predictable way, with angular distance from the zenith and is brightest near the horizon.

Some concept of these levels of night-sky brightness can be obtained from the descriptions of the sky given below. The sky under natural conditions alone is crowded with stars, extending to the horizon in all directions. In the absence of haze the Milky Way can be seen to the horizon. Clouds appear as black silhouettes against the sky and stars look large and close. About 2500 stars can be seen across the entire sky.

At a brightness level of 1.1 times the natural level (a 10 percent addition of man-made illumination), the sky appears essentially as above, but a glow in the direction of one or more cities may be seen near the horizon. Clouds are bright near the city glow. With a 50 percent increase the Milky Way remains brilliant overhead but cannot be seen near the horizon. Clouds have a grayish glow at the zenith and appear bright in the direction of city glows.

When the sky is twice the natural light level, as it may be in thinly populated suburbs or small towns, the Milky Way is magnificent but contrast is markedly reduced and delicate detail is lost. The limiting magnitude (magnitude of the faintest star visible) is noticeably reduced. Clouds are bright against the zenith sky. Stars no longer appear large and close. At a brightness level of five, typical of a suburb, the Milky Way is still visible overhead but not near the horizon. Faint stars are reduced in number near the horizon. The sky has some color away from the zenith.

At ten times the natural amount, the Milky Way is only marginally visible and only near the zenith. The sky is bright and discolored near the horizon in the direction of cities. The sky looks dull gray in color. In the city a 25-fold increase in light is commonplace. Stars appear weak and washed out and reduced to a few dozen. The sky is bright and discolored everywhere.

One point of possible confusion may need explanation here. It has been stated several times that near the horizon fewer stars are visible due to haze. This is not strictly true. If we think of the Earth's atmosphere as a distinct layer of clear air surrounding the planet with a uniform thickness and clarity, it becomes apparent that the obscuration due to clear air is minimal in the direction toward the zenith and increases with angular distance from it. Astronomers define the atmosphere in the direction of the zenith as of unit thickness, called an air mass of one, or one unit of atmosphere. The air mass at any other direction is a function of the angular distance from the zenith called *z* (see the figure on this page). At a zenith distance of 60 degrees, equivalent to an altitude of 30 degrees above the horizon, the air mass becomes equal to twice that at the zenith itself, thus we are looking through twice as much atmosphere in that direction. As we consider directions closer to the horizon, the air mass grows quickly. We are looking through three air masses at an altitude of about 20 degrees and at 15 degrees starlight must pass through four.

It has long been known by how much an atmosphere composed of clear air free of clouds, dust, and haze dims the incoming starlight at the surface. We are aware that at sea level we see stars about 0.3 magnitudes fainter at the zenith than we would if

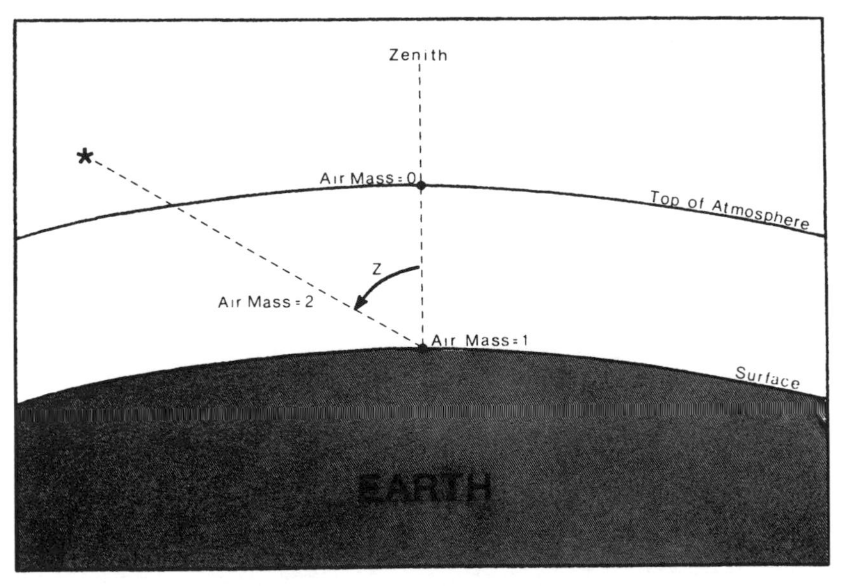

Light from a star passes through twice as much air (an air mass of 2) at an angle of 60 degrees from the zenith as it would if at the zenith (an air mass of 1).

the atmosphere didn't exist or if we were above it. At two air masses the diminution would total 0.6 magnitudes. Stated another way, the stars would appear 0.3 magnitudes brighter overhead if, like the Moon, we had no air.

The dimming of light by a specific amount per air mass is called the extinction, and the amount, the extinction coefficient. The diminution is smaller than 0.3 magnitudes per air mass in a very dry climate or at an elevation considerably above sea level. At some of the large mountaintop observatories in the southwestern United States it may be as small as 0.15 magnitudes.

An extinction of 0.3 means that on a night where, typically, the faintest star visible at the zenith is near magnitude 5.5, the limiting magnitudes must brighten with increasing air mass, and at 30 or 20 degrees above the horizon (at two or three air masses, respectively) the faintest visible stars will be of magnitude 5.2 and 4.9, respectively. It is evident that near the horizon, the stars appear to fade out even on the best of nights.

A hazy night or one at a site with considerable light pollution will show an extinction well in excess of 0.3 magnitudes. In a suburban neighborhood, a coefficient of 0.5 is very common even on the clearest of nights. This means that at the zenith and at two and three air masses, the limiting magnitudes will be near 5.3, 4.8, and 4.3, respectively. In the suburbs the sky washes out with approach to the horizon far more quickly than it does at a darker site and is much more noticeable than the difference between the two sites in the appearance of the stars overhead.

The eye does not respond in a linear manner with increasing light intensity because its light detection cells in the retina, the light-sensitive surface at the back of the eye, are of two kinds, the rods and the cones. Each serves a specific purpose. The rods are related to scotopic vision, the night vision that allows for great visual acuity. The cones predominate at the greater light levels of daylight, providing photopic vision with its emphasis on color sensitivity and rendition.

At the center of vision, or fovea, visual acuity is at its greatest, but there the cones are most concentrated for maximum detail, and there are no rods at all. For this reason, night-sky observers are advised to use averted vision when looking at faint light sources such as stars. At a distance of just a few degrees from the fovea and beyond, the rods predominate, and much fainter stars can be detected than at the fovea itself.

The changeover from scotopic to photopic vision as light sources increase in brightness affects the limiting magnitude of the faintest star one can see. This changeover point is not known precisely, and the limiting magnitude of the faintest visible star may not be a fully satisfactory method for the determination of sky brightness.

Other more reliable means are readily available for that measure. One of the most efficient involves the photography of the zenith night sky under strict constraints of lens opening, film speed, and exposure times. It was first developed by astronomers at the Tokyo Observatory and is now in worldwide use. Careful measures reveal that most night-sky brightness due to man-made illumination is caused by lights within only a few miles or kilometers of the site of observation. Light pollution control is most important within the immediate neighborhood of an observer. Sometimes an adequate view of the night sky requires a move of only a few blocks.

When a person enters a darkened room from brighter surroundings, the pupil widens quickly to typically about seven millimeters, or to about one-third inch from about one-half to one-fourth this diameter. Much more light enters the eye to compensate for the darkness, but full dark adaptation takes another 20 to 30 minutes. All of us have experienced this after entering a movie theater and trying to spot empty seats. Later we can see easily, due to visual purple, a substance that builds within the eye and greatly increases its sensitivity. Night-sky observation also requires about this time interval for best viewing.

The diameter of the dark-adapted pupil is, as noted, about seven millimeters in a young person and reduces to about five with age. This directly influences the choice and use of binoculars. Field or opera glasses and binoculars are specified by two numbers; the commonest among the latter are 7 by 35 or 7 by 50. The first of these, usually 7 but sometimes 10 or even 20, refers to the magnification; objects are seen 7 times larger and closer than with the naked eye. The second number refers to the diameter in millimeters of the objective lenses, the large ones away from the eyes. At 25.4 mm to the inch, 7 by 50 means that the large lenses are just about two inches across. Opera glasses magnify an image only about three times and serve an entirely different purpose. They are not customarily recommended for night observation outdoors.

If the first number is divided into the second, the result is a measure of the diameter of the exit pupil, the circle within which the light exits the eyepieces. Binoculars of 7 by 35 and 7 by 50 have exit pupils of five and about seven millimeters, respectively. The former size is more compact and easier to handle and carry, but the latter size is recommended for night work because its exit pupil is at least as large as the pupil of a dark-adapted eye and the full gain in brightness is obtained. Powers higher than about seven are not recommended for use without a tripod. The greater magnification results in unacceptable jiggling because the hands cannot hold binoculars completely still.

Although telescopes of greater magnifying power reveal many more celestial features, binoculars form a very valuable first step in optical aid.

Chapter 12

The Downside of Light

We grow accustomed to the dark, when light is put away,
As when the neighbor holds the lamp, to witness her Goodbye
A moment—We uncertain step, for newness of the night—then
Fit our vision to the dark, and meet the road erect
Either the darkness alters, or something in the sight adjusts itself to midnight, and life steps almost straight.

—Emily Dickinson

Can we ever grow accustomed to midnight and the dark? Percival Lowell, scion of a family of Boston Brahmins, is buried alongside the observatory he founded near Flagstaff, Arizona. The epitaph on his tomb tells us that "he loved the stars too fondly to be fearful of the night." All my life I have been something of a nocturnal creature. I was among those who welcomed the two bright comets that have graced our sky in the last few years. They helped show millions that darkness in the sky has a rightful place.

Our species receives most of its stimuli through the visual sense. Whenever that sense was unable to protect our kind from predators real and imagined, we cringed; we imagined things that weren't there and much in our culture still demands that darkness be shunned at all costs. Vampires, poltergeists, wraiths, and phantoms persist in our imaginations, and nyctophobia, the fear of darkness, is America's fastest growing phobia. If my students form an unbiased sample, few among Generation X have ever seen the Milky Way; their galaxy is taken from Star Wars and Star Trek. The cyberworld may be accessible, but a sense of wonder at the night sky is not.

Proposals to NASA or some foreign government have been made to launch a satellite whose purpose is advertisement. We have the technology to send up a space satellite into orbit that would then unfurl many square miles of tough thin fabric such as mylar. It would likely be a billboard in the form of a giant sail that would appear larger and brighter than the full Moon in the hours after sunset and before dawn. If any of the several proposed advertising satellites were now in orbit and lighting up the night sky, pushing a soft drink, a cigarette, or one more way of building up one's "abs," even bright comets and their tails would be mostly invisible. In time such a satellite display would be broken up under the bombardment of micrometeorites into debris that could spread itself all around the satellite's orbit, with no way to clean it up. Until now, public outrage has spared us appalling debacles like these, but no law prevents this or any other nation from similar actions in the future.

Preservation of darkness requires active effort, propelled by the belief that darkness is deserving of it. This century closes on a multiplicity of sprawling cities with well over a hundred of them above two million in population, as compared with four in 1900. Urbanization and suburban sprawl, coupled as they are with electrification around the world, have led to an explosion of light, to the extent that the glow over the dark side of the Earth is now visible throughout much of the solar system to telescopes like ours.

Some might feel that bringing the rapid growth of light pollution under control must occur at the national or international level of legislation and treaty. But the NASA reproductions in Chapter 11 indicate that this is not necessarily so. The view of eastern North America intimates that the entire urban corridor extending from Boston through Washington, for example, is awash in light. But a contradictory picture is revealed by the image of the region surrounding New York City at night. Even on Manhattan Island, the illumination comes not in the main from the gaudy lights of Broadway and Times Square, nor is it a product of the great skyscrapers. Instead, as noted earlier, the pervasive orange color of the original image reveals that ordinary sodium vapor streetlighting accounts for most of the pollution. The difference between the two images lies in their resolution. A close examination of the latter reveals a number of relatively dark regions in the suburbs, if not in the city. The communities in those regions can do much through ordinances at the local level. Amateur astronomers, birders, and others can unite behind a full-cutoff shielding provision that will darken the sky by more than half to everyone's benefit. Awareness is the first part of the solution.

Emily Dickinson knew and paid tribute to the darkness of the night sky. What would she have thought of the Washington Monument and its nighttime illumination? We and she might countenance the illumination of one monument—one of a kind. But the garish sky above the monument shown in the figure

The Washington Monument at night illuminated by spotlights directed upward.

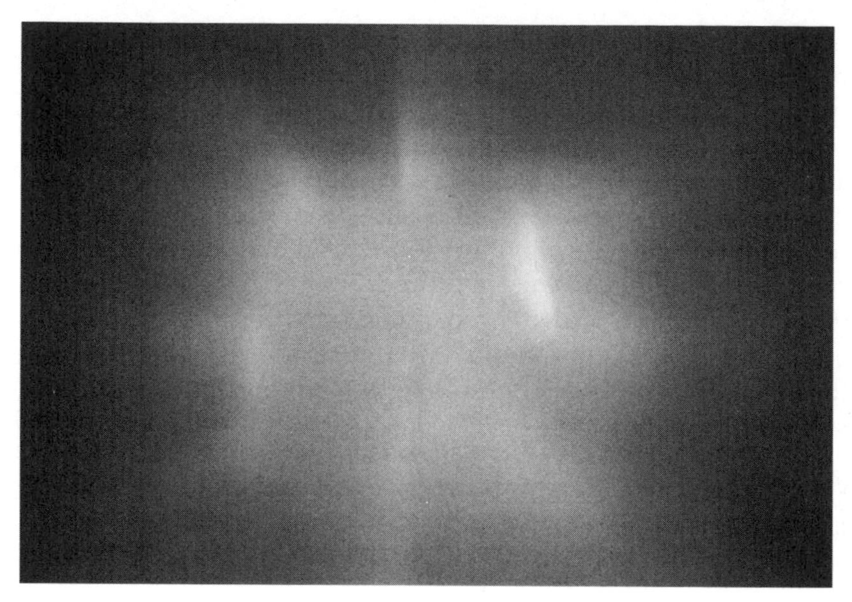

The sky directly above the Washington Monument, with an overcast ceiling of 2000 to 3000 feet showing the glare from the spotlights. This photograph was taken directly after the one on page 115.

on this page is an ingredient of every one of the millions of billboards lit every hour of the night from below. Their excess light, too, spills heavenward, all night, every night. It is time to shield lights from above the horizontal plane, at the minimum; after a few years of cost recovery, we save money and energy, and everyone wins. Only in this way can our society create the kind of social buffer that could mount a lasting resistance to any threat from an advertising satellite. We are well on the way to that kind of social opposition to contamination from secondhand smoking of tobacco; let's do it now with costly, wasteful secondhand light.

CHAPTER 13

The Moon

How sweet the moonlight sleeps upon this bank!
Here will we sit and let the sounds of music creep in our ears.

—WILLIAM SHAKESPEARE
The Merchant of Venice

The Moon is probably the most gazed-upon object anywhere, on or off the Earth. It is visible from all parts of the Earth; everyone can see it. The Sun is usually too bright to view directly with comfort and safety, and no one planet or star is anywhere near bright enough to attract attention as does the Moon.

Along with the fascination it holds for the artist, the poet, the stargazer, the professional astronomer, and the geologist, the Moon has powers both real and imagined. It influences the mind and body, it brings about eclipses and tides, and it gives rise to weather lore and various spiritual and paranormal activities. The word *lunatic* derives from the Roman goddess Luna, the Latin name for the Moon. The werewolf is considered one of many creatures under its influence. It has inspired science fiction and space exploration. Its proximity, physical structure, and gravitational influence have affected the development of the Earth and life on it in many ways, all compelling objects of study.

Certainly the most noticeable property of this companion of the Earth is the set of phases that it displays each month. The

Moon is a cold, lifeless world that gives off no light of its own. It appears bright because it reflects the Sun's rays. Since its period of rotation is the same as its revolution about the Earth, it keeps the same face toward us. But the side facing the Sun changes continually, and the phases occur as the Moon presents more or less of the bright hemisphere as it moves about the Earth. The Moon appears brilliant in the dark night sky, especially near the full phase, but in reality it reflects less than ten percent of the incident sunlight, about as much as the rocky material on the Earth.

The figure on this page shows how the phases come about. When the Moon is in position 1, it is invisible to us for two related reasons. Its dark side is turned toward us, the bright side is toward the Sun, and in addition it appears so close to the Sun in the sky that it is lost in the solar glare. At position 2, the Moon has swung around in its orbit so that we see a bit of its bright side and it appears to us as a crescent. At position 3 we see half

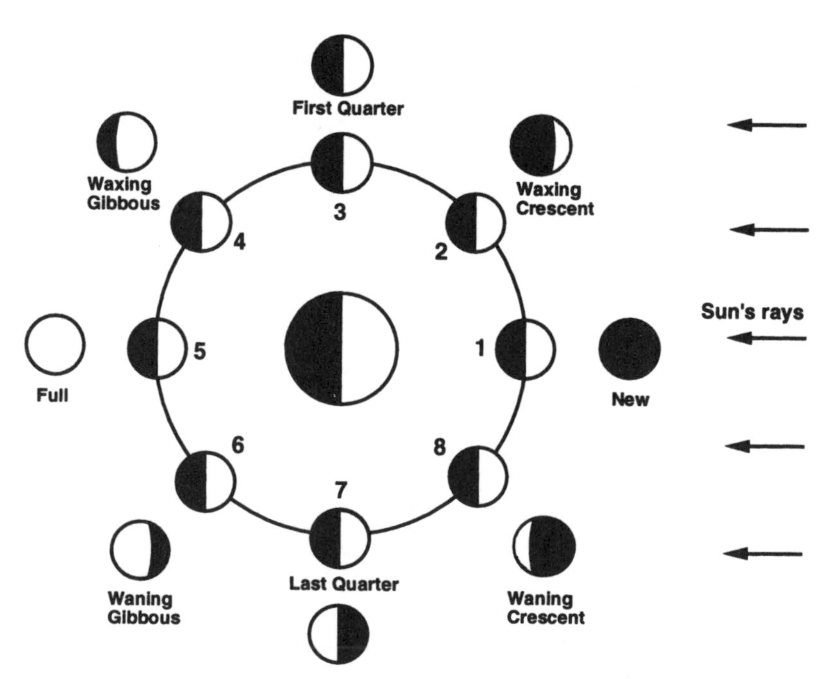

The phases of the Moon. The outer ring of moons shows the phases for each position of the inner ring.

the Moon illuminated. We speak of it as at the first quarter because it has gone one-quarter of the way around in its orbit. Position 4, when it appears more than half lit but yet not full and round, is called the gibbous phase. This is followed by the full Moon at position 5, which, strictly speaking, lasts but an instant when the Moon is opposite the Sun. Most people describe the Moon as full for a few days surrounding this moment when it looks nearly round in shape. The full Moon appears to last in a sense for no more than about one-eighth of the entire month. This apparent length may account for the perception that a large number of bizarre events and displays of criminal or erratic behavior occur during the full Moon. Objective studies of this effect, which must take these perceptions into account, have not revealed an increase in lunacy or crime with the period around the full Moon.

After the Moon is full, it travels through positions 6, 7, and 8, showing more and more of its dark side, until position 1 is reached, when the sequence is begun all over again. At position 7 the Moon is again half illuminated; this phase is known as last, or third, quarter because the Moon has traveled three-fourths of the way around since the last new Moon. From new to full, the Moon is waxing, and from full back to new, it is waning.

One entire set of phases from one new Moon to the next lasts $29\frac{1}{2}$ days and is known as a lunation. The count of days always starts at new Moon, thus at the first-quarter, full, and last-quarter phases, we sometimes speak of a 7-, 15-, or 22-day old Moon, respectively. Some early peoples actually believed that a brand new Moon was created with each new phase, and upon reaching the full phase, the old Moon was devoured by mice or a dragon, depending on the culture, until it was gone. The name new Moon may have stemmed from that belief.

Other people expressed the belief that the old Moon gradually burned out and faded away, since the entire Moon can occasionally be seen faintly after new when the bright waxing crescent is seen. The correct reason for the occasional visibility of the entire Moon is that the faintly lit portion is being illuminated by earthlight, light that has been reflected by the Earth onto the lunar surface, brightening it to the point where we can see it faintly. Seen from the Moon, the Earth would appear much big-

The crescent Moon showing the entire lunar surface illuminated by earthshine. The much brighter sunlit crescent is necessarily overexposed. Courtesy of Yerkes Observatory.

ger and brighter than the Moon does to us, as we shall see in the next chapter.

If you wish to become familiar with its appearance and motion, it is best to observe the Moon over a period of time. After you go out and look at it for several nights, you will be better able to visualize the relationships between Sun, Earth, and Moon, what exactly causes the phases of the Moon, and how they are closely linked to the times of moonrise and moonset. Observation makes it obvious that the Moon shines only by reflected light from the Sun.

The best time to begin nightly observations is just after the new Moon. At this time the Moon will be visible in the evening sky shortly after sunset every clear night during a two-week period. It will first be seen in the west toward the brightest part of the sky at twilight and move toward the east among the stars by about 13 degrees (about the width of your hand at arm's length) each successive night. Two weeks later it will be full, and soon it will not be visible at dusk, but can be seen to rise later on, still before midnight. Then for the last week of its lunation, it will not rise until the morning hours. The Moon can, of course, often be seen during the daytime when the sky is clear. Once you have a feel for where the Moon will be seen at dusk, you may be able to relate a daytime sighting to another after sunset.

As noted earlier, the duration of one lunation lasts $29\frac{1}{2}$ days. But the true orbital period of the Moon about the Earth, called the sidereal period, is only $27\frac{1}{3}$ days. The longer period is the one we see and notice, and the one that defines the month in our calendars. The difference accrues from the fact that the Earth moves about the Sun carrying the Moon along with it. If, for instance, a star were to appear directly above the full Moon, then $27\frac{1}{3}$ days later the Moon would again pass the star but it would not yet be full. In the intervening period the Sun has appeared to proceed along the ecliptic almost one twelfth of the way around the sky. It takes the Moon another two days to catch up to the Sun again; thus the phase month is the longer period.

Two other well-known features of the Moon deserve comment. The Moon appears larger near the horizon than it does farther aloft. Although no one knows for sure why this is the case, we can venture some guesses about it. First, every celestial object appears larger near the horizon, including the Sun and constellations as well. The Big Dipper stretches on forever in the fall when we see it below the pole star not far above the northern horizon, but in the spring when high in the sky it is seen to be no larger than many other star groups.

Some have speculated that the horizon is often littered with trees and buildings behind which the Moon seems to loom. But this doesn't explain the fact that this enlargement is also seen on

A typical drawing of the rising full Moon. The Moon's size is often greatly exaggerated due to the impression that it is very large when seen near the horizon.

the open sea where the horizon is an unobstructed line between sea and sky. The Moon appears larger there as well, but maybe not as much larger as it does on land.

Even if you cannot afford an ocean cruise to some exotic port of call, you can test this speculation with a cardboard tube from a roll of paper towels. If you view the Moon through such a tube, does it appear as much larger near the unobstructed horizon as it does when viewed near trees? Try looking at the Moon with and without the tube when it is low and high in the sky. See if you can get a feel for the relative sizes in the various locations.

A substantial variation in the apparent angular size of the Moon does occur. The Moon's orbit, like those of the planets, is an ellipse, with an eccentricity of 5.5 percent. At perigee, when the Moon is closest to us, it appears some 11 percent bigger than

The extreme variation in the angular size of the Moon, varying from largest at perigee to smallest at apogee. Courtesy of Yerkes Observatory.

it appears at apogee, at its farthest point from us. This apparent difference in size is not related to the angle of the Moon above the horizon.

The second commonly noticed lunar feature is the flattening that the Moon seems to undergo whenever it is very near the horizon. The figure on page 124 shows the effect on the full Moon just rising over Denver, Colorado. The effect is more evident in the case of the Sun, since it is so much brighter than the Moon and is more often visible when just above the horizon. Both seem oblate, not round, much as Jupiter always appears. But in Jupiter's case, the oblateness is real because that bulky planet rotates

Full Moon just rising over Denver, Colorado. It appears oblate due to differential atmospheric refraction.

on its axis in only ten hours, giving it a very substantial equatorial bulge. Its equatorial diameter is 1 very conspicuous part in 15 larger than that between its poles; for the Earth the ratio is 1 part in 300—not obvious at all.

Both Sun and Moon rotate slowly and are more truly spherical than the Earth. The cause of their apparent oblateness is refraction due to our atmosphere. Light from the zenith is unaffected, but an apparent shift in position upward of their disks increases with decreasing angle above the horizon. At an angle of 45 degrees, the shift amounts to only about one minute of arc (about one thirtieth of the Moon's apparent diameter). But close to the horizon we are looking through so much more air that refraction adds up, and is even differential, that is, noticeably variable across the lunar and solar disks. The lower limb, or edge, of either disk when seen just on the true horizon; that is, 90 degrees from the zenith, is raised by 0.6 degrees, whereas the upper limb is raised by only 0.5 degrees, making the Moon and

Sun appear only 0.4 degrees in height, while retaining its customary width of half a degree. In other words, the lower limb of the full Moon is refracted upward more than the upper limb because we are looking through more air to see the lower limb, thus squashing the Moon into an elliptical shape.

The atmospheric refraction is such that at the moment when the whole disk can just be seen, its true position is just *below* the horizon. Sunrise and moonrise appear to be about two minutes earlier than they really are, and sunset and moonset, the same amount of time later. Daylight would be shorter by some four minutes or more if the Earth had no atmosphere.

* * *

In art and literature, the Moon appears in many guises. In most cases it is portrayed correctly, but the largest group of astronomical mistakes probably involves an incorrect depiction of the lunar phase with its orientation in the sky. Common errors include the portrayal of the full Moon in the daytime sky. Since the Moon at full must lie opposite the Sun, both cannot be well above the horizon at the same time. Furthermore, the horns or points of the crescent Moon must always point away from the Sun. Thus they never point downward at night.

For all the renderings of the Moon at full, or even as a crescent, artists have only rarely portrayed it in its gibbous phase. One who did was Jean-François Millet, the mid-nineteenth-century French painter whose *Sheepfold by Moonlight* depicts a waning gibbous Moon just rising over the eastern horizon. I use this painting as a test for students in my introductory courses in astronomy of their knowledge of the Moon's phases and their relation to the time of day or night. Assuming that Millet painted in France, we can set the time of the scene at middle or late evening, and probably in the winter.

The Moon is seen, by day and by night, more frequently at the gibbous phase than at any other. At first this fact may seem improbable; surely the cute little crescent and the romantic round full Moon must be as commonly seen as is that asymmetric football. But for how many days in each lunation do we judge the Moon to be full? When asked to call its phase, people refer to it as full for no more than about two days on either side of the exact moment of its opposition to the Sun in the sky, or four days in all. And at either quarter, it is seen as a half Moon for a day or so at most, for the terminator, that line between light and dark on the lunar surface, appears as a straight line only briefly between the concave crescent and the convex gibbous phases. The crescent lasts as long as the gibbous, but when it is thin, near the new moon, we can barely see it and then only in ideal conditions just before dawn or after dusk, and in the daytime it is just too faint and lost in the solar glare. But the football-shaped moon is bright and easily visible in a clear sky in the daytime and at night.

* * *

A star cannot lie directly between the points of the crescent Moon because the entire lunar disk is always present even if part of it is dark. In the *Rime of the Ancient Mariner*, Coleridge describes the "horned Moon with one bright star within the nether tip." This has often been cited as a mistake, but Coleridge was very observant, and because no other error appears among his many allusions to astronomical effects in the poem, he is not likely to have made this one. We can assume his star is enough off center to avoid being hidden by the invisible lunar limb.

Several scholars have analyzed and annotated Coleridge's astronomy and conclude from his notes and the scientific knowledge of his times that he meant to attach an extraordinary atmosphere to the narrative if not a supernatural one.

Insight into the Moon's phases is also revealed by William Shakespeare in the opening lines of his *Midsummer Night's Dream*:

> THESEUS:
>
> Now, fair Hippolyta, our nuptial hour
> Draws on apace. Four happy days bring in
> Another moon; but, O, methinks, how slow
> This old moon wanes! She lingers my desires,
> Like a stepdame, or a dowager,
> Long withering out a young man's revenue.

HIPPOLYTA:

Four days will quickly steep themselves in night,
Four nights will quickly dream away the time;
And then the moon, like to a silver bow
New-bent in heaven, shall behold the night
Of our solemnities.

Nowhere do literature and the arts overlap with astronomy more than in the observation and study of the Moon. In many cases the artistic or poetic license taken is too great to match the observed Moon as it is really seen, but in this instance the Bard seems to have made his case with the real Moon in mind. The midsummer night from which the play takes its title occurs most likely on the eve of June 24, known as Saint John's Eve and widely celebrated throughout Europe. We know the phase and the appearance of the Moon when Theseus makes his remarks and how they will vary over the four days of the action of the play. The phase must be a waning crescent, because in four days' time the Moon will realign with the Sun at new Moon and will not be visible. This interpretation makes more sense if "now" is substituted for "new" in Hippolyta's speech, as some commentators have suggested.

* * *

The Moon has had a major place in the other arts as well. Some music calls out for a ripe full Moon gleaming over a pastoral countryside. The long, exuberant theme in the final movement of Rachmaninoff's Second Piano Concerto is one example; perhaps this is forecast by the lyrics later set to it by Buddy Kane and Ted Mosman and popularized by Frank Sinatra: "Full Moon and Empty Arms." Henry Mancini's "Moon River" and Glenn Miller's popular "Moonlight Serenade" might well be others. "Clair de Lune" by Debussy does not expressly mention a lunar phase, but somehow we all know that he had a full Moon in mind. Beethoven's "Piano Sonata Op. 27, No. 2, Quasi una Fantasia," which wasn't dubbed the Moonlight Sonata until many years after it was written, seems also to shine by a full Moon.

Not all night music is so endowed. For the nocturnal urban music of Copland's "Quiet City," or the middle movement of the Piano Concerto in F of Gershwin, the full Moon does not seem appropriate; perhaps a crescent moon hanging in a murky sky, just over a cityscape of rooftops and water tanks, might do. Still other night music seems moonless altogether. The middle movements of Bartok's piano concerti, for example, are evocations of the ambience of a moonless night, as is the Serenade for Tenor, Horn and Strings, Benjamin Britten's testament to the world of darkness. The menacing side of night and darkness has its representatives too, in "Danse Macabre" by Camille Saint-Saëns, in Modeste Mussorgsky's "Night on Bald Mountain," and, above all, in Sir Andrew Lloyd Webber's musical resurrection of the *Phantom of the Opera* and his paean to the "Music of the Night."

"When the Moon hits your eye like a big pizza pie . . ." begin the lyrics of a song popularized by Dean Martin. Just when might the Moon resemble a pizza? When it is big, round, reddish, and near the horizon.

We've all seen the full Moon as depicted in countless horror movies. From the original 1931 film on, Dr. Frankenstein and his ungainly monster, so well played by Boris Karloff, seemed always to appear against a backdrop of moonlight. The Moon was always full and viewed inevitably through broken clouds and haze. Twelve years later the monster encountered the Wolf Man in another film. Lon Chaney, Jr., as the Wolf Man, was a normal person until the full Moon rose; then its rays converted him into his hirsute alter ego. In a spate of artistic license, the full Moon was made to rise and rise again, night after night for a fair portion of the month, always full and always through the same veil of clouds and haze. Perhaps the cloudy and rainy nights and nights with little or no moonlight are reserved for film noir, which is, of necessity, darkly illuminated. In one of the best-known of these movies, *The Third Man*, the celebrated first appearance of Orson Welles standing in a shadowed doorway is unimaginable in bright light, whether that of the Moon or of glary modern streetlights.

What would have been the destiny of the horror movie genre if we had a moon or moons like those of Mars? That planet's two tiny moons, neither bigger than a mountain, are too small to be round. At something like ten and five miles in diameter, Phobos and Deimos are potato-shaped blobs, as is any celestial object less than about ten times their size. The reason for the spherical shape of larger bodies is that they contain enough mass for their gravitational fields to overcome the natural rigidity of the rocky or icy material of which they are made and render them globular. Perhaps Mars has no werewolves or vampires because Phobos would appear irregular and lumpy, and the smaller, more distant Deimos, as just a speck of light in Martian skies. Mercury and Venus possess no lycanthropes either, if only because they have no moons at all. And who wants to rhyme a potato with June or a sleepy lagoon!

A flight to the Moon, whether or not on gossamer wings, as Cole Porter once remarked in "Just One of Those Things," is a major task. Before the first manned landing on July 20, 1969, some of us had been imagining such a trip for many years. In the 1902 film, *Voyage to the Moon,* adapted by G. Méliès from the book by Jules Verne, the man in the Moon is shown with the spacecraft lodged in one eye. Interesting indeed is a comparison between the thoughtfully made 1950 movie *Destination Moon,* and the recent *Apollo 13,* recounting the near-disastrous flight of that name in 1970. We see the earlier effort now as a period piece with dated sets, but all in all it gave a tolerably realistic account of a first Moon voyage, in the technical details if not in the plot. Perhaps the most erroneous and dated evidence shared by pre-Apollo pictures and film sets is the portrayal of the world as all but cloudless. Once it was imaged from space, we realized that the Earth is about half enshrouded in clouds at any one time.

* * *

The Moon has played a large role in weather lore ever since people discovered the possible association between the two. The Moon's unique place in weather lore is primarily due to its brightness in the night sky. Today we know that the Moon's actual effect on the weather may be best described by a bit of

The Moon is drawn whimsically to show the landing of a spacecraft in an early film.

weather wisdom appearing in Arthur Machen's *Notes and Queries* of 1882. He attests that

> The Moon and the weather may change together;
> But change of the Moon does not change the weather.
> If we'd no Moon at all, and that would seem strange,
> We still should have weather that's subject to change.

There are many places on the Earth where the wet and dry or hot and cold seasons coincide with extremes in the orientation of the ecliptic in the sky and consequently of the Moon's appearance as well. But no causal relationship has been established, and it is highly unlikely that one exists. If these orientation extremes happen to correlate with seasonal climatic effects, as they often do, they only confirm Machen's first line, not a case for a causal relationship between Moon and weather. The horns of the waxing crescent Moon in the evening sky after sunset point up or sideways depending on one's latitude and the time of year and not much else.

Some phenomena do correlate with weather changes; everyone has heard that a ring around the Moon is a harbinger of foul weather. That old belief deserves some clarification. First, two kinds of rings can be seen to surround the Moon at night or the Sun in the daytime. One is a wide ring of about 23 degrees in radius that can fill a fair portion of the sky if seen completely around the Moon; this type of ring, called a halo, is caused by moonlight passing through ice crystals usually high above the ground near the stratosphere, a phenomenon associated with cirrus clouds, those little feathery clouds that are too thin to block light from the Sun or Moon. Cirrus clouds, especially if seen to thicken, may be associated with an approaching front containing heavier clouds bearing rain or snow. A halo is not always indicative of precipitation; however, and is not a very trustworthy omen. Nevertheless, if it becomes more intense as the night wears on, it is of some use as a sign of foul weather. The other type of ring is called a corona and is brightest right at the limb of the Moon (or Sun), fading away a degree or two outward. A corona indicates a lower cloud composed of water droplets. Its value in predicting inclement weather is about as good as the halo. Occasionally a bright planet such as Venus or Jupiter can be seen enshrouded in mist or a corona, but stars are generally just too faint to illuminate a halo or corona.

Other moon signs have been used to predict the weather; hence, Clear Moon, Frost Soon, and A Pale Moon Doth Rain.

The second adage falls into the same divining group as coronae and halos: If the Moon gets paler, it is due to clouds that may be on the increase. The first saying is often true because on very clear nights the cooling of the Earth's surface is greatest, and the drop in the temperature can bring on the condensation on plants and lawns that we know as frost or dew, depending on whether the temperature does or does not reach the freezing point.

Finally, here is an old saying from Sir Patrick Spens about the appearance of the Moon at the crescent phase:

> I saw the new moon late yestreen
> With the old moon in its arm.
> If ye be goin to sea, sir
> I fear you'll come to harm.

The old moon is another name for earthshine caused by the double reflection of sunlight off the Earth to our west, which is still in daylight, and then off the Moon and back to us. There can be no connection between the visibility and brightness of this feature and the chances of a safe voyage.

* * *

One of the best-known effects of the Moon is on the tides. Anyone who has sailed on the ocean or spent a day at the beach knows of their existence, and that two high tides and two low tides occur at any given spot each day (except in rare cases where the coastline is irregular in some manner). But other features about the tides are not as obvious, and more careful and patient observations are needed to notice them.

It should come as no surprise that Isaac Newton was the first to correctly explain the cause of the tides since they result from the gravitational force of one astronomical object on another. This means that (1) the masses of the Earth and the Moon and (2) the distance between them govern the force of the tide raised. However, there is one very important difference between the tide-raising force and the force of gravitation that relies on these same two factors. The tide-raising force, or tidal force, is a *differential* force, not a fixed, or absolute, one, as is the gravitational force. In the case of the Moon, for example, the size of this tidal force is a function of the *difference* between the point of interest and a reference point, usually taken as the center of the Earth. The figure on page 133 shows the situation in which three points are

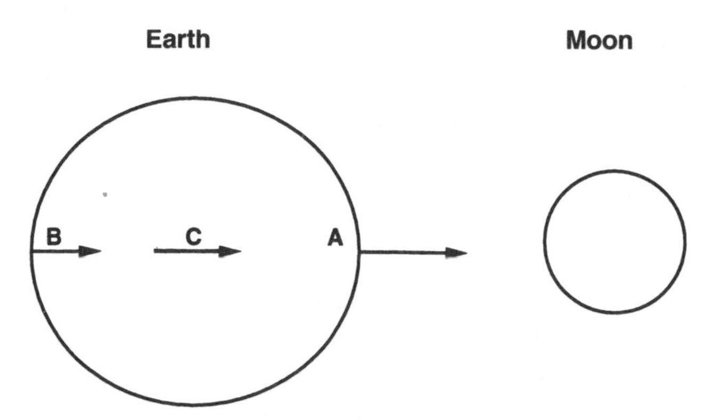

The tides result from a differential gravitational force from the Moon. The force is greatest at A, the closest point on the Earth's surface to the Moon, and least at B, the farthest point, with the force at the center, C, in between. The Moon pulls on the side nearest it more strongly than the farthest point.

considered. Point A is the sublunar point, that point on the Earth that at the moment sees the Moon at its zenith. Point B lies at its antipodes, the point opposite A on the sphere. The center of the Earth is represented by C. Most people understand the bulge in the direction of the Moon, but the fact that there is another in the opposite direction is a more difficult concept to explain that I will leave for another time. The tidal force turns out to be an approximate function of the inverse cube, or third power, of the distance between Earth and the Moon, whereas the well-known gravitational force varies as the inverse square of the distance between the two.

Not one but two objects give rise to the tides. These are the Moon, because it is so close, and the Sun, because it is so big. The planets raise our tides too, but they are neither so close nor so big and their effect can be neglected in all but the most precise work. The ratio between the forces due to the Moon and Sun are about 2.2 to 1 on average; that is, the Moon raises the oceans a little more than twice as much as does the Sun.

If the Sun affects the tides almost half as much as does the Moon, and the tidal bulges are oriented along the line from the Earth to the tide-raising body, how do we account for the effects of different orientations of the Sun and Moon? The difference in alignment varies, of course, being very small at new or full Moon and largest at each of the quarter phases. Whenever the three bodies are in near alignment, as they are at either new or full Moon, the arrangement is called a *syzygy* (a word that cannot be spelled in Scrabble without a blank). The quarter phases are sometimes known as quadratures. At a syzygy, the tidal forces of the two bodies reinforce each other, resulting in tidal extremes with very high and very low tides and a larger swing or amplitude between them, a condition known as a spring tide. At quadrature, when the tidal forces counteract each other, the bulges align with the stronger lunar tidal power but not by as much; thus the high and low tides are not as high and low. This is the neap tide condition.

Thus we have three pairs of words that apply to the tides: high and low, spring and neap, and also flood and ebb. Flood tide refers to the period when the tide rises from low tide—when the tide is "coming in." Ebb tide refers to the period from high to low tide when the tide is "going out." Since the tide goes with the Moon regardless of its phase, the period between one high tide and the next is not half a day, but a little longer. Because the Moon orbits the Earth in about 27 days, in any one day it moves about $\frac{1}{27}$ of the 360-degree revolution it must make each month. Then it takes about 24 hours and 50 minutes of time to return to the same point on the average. The time between successive high tides is about half this, or 12 hours and 25 minutes; thus 6 hours and 12 minutes of flood tide are followed by the same period of ebb tide.

The interval between one spring tide condition and the next—between one syzygy and the next—has to be half a month or some two weeks. The neap tides fall in between, bringing about lapses of about one week from spring to neap tide or vice versa. Along with this variation is one due to the ellipticity of the orbits of the Moon about the Earth and the Earth about the Sun. The distances from the Earth to both Moon and Sun vary such that the ratio of their distances can range from about 360 at apogee in January to about 420 at perigee in July, with an average near 390. Because of this variation, the average tidal force ratio of 2.2 to 1 given earlier actually goes through an annual minimum of about 1.7 to 1 and a maximum of nearly 2.8 to 1. This surely affects the magnitudes of the spring and neap tides actually observed over the year.

If one measures the tide along a shoreline, one will quickly find that local conditions impose even more variations on this pattern. These are due to coastal irregularities and varying ocean depths along the nearby continental shelf. Most noticeable is the varying amplitude between high and low tides. On the open sea, it averages about half a meter or 20 inches (more and less for spring and neap conditions). But if the nearby ocean floor and coastline act as a funnel, it can be much higher. The tidal championship goes to the Bay of Fundy between the Canadian maritime provinces of New Brunswick and Nova Scotia, where the mean swing is a whopping 53 feet. If the funnel is a river, the tidal bore, as the flood tide is called, can back up the normal flow for a long way. At Albany, New York, for instance, the tidal range is six inches, even though it lies some 140 miles upstream on the Hudson River from its mouth in New York City. And at Saint John, New Brunswick, at the mouth of the river of the same name, the tidal bore reverses the falls there. The water spills one way during flood tide and the other way during ebb tide. Even the Great Lakes are large enough to have measurable tides; at Chicago a range of an inch or two can be detected.

The times of high and low tides also differ widely from one port to another. Neither they nor the amplitudes can be predicted from theory at any port. There are just too many unknown factors involved. The "establishment of the port," by which times and heights of tidal extremes can be calculated, is derived from observations alone. The establishment is not just an idle exercise. At some ports the larger ships can only pass safely over sand bars during or near high tide lest they be grounded.

Credit for the establishment of ports is sometimes given to the Venerable Bede (A.D. 672–735), a monk who lived in northern England. This remarkable polymath is best known for his lucid historical works, but he was an astronomer seemingly without peer throughout the whole of Europe.

A few final remarks about the tides. We think of them as related only to the levels of oceans and large lakes, but the water component of the Earth is not the only part affected. Being gaseous and less rigid than water, our atmosphere is also pulled about by the lunar and solar tidal forces. We don't notice it because we live at the bottom of the atmosphere, just as the ocean tides would not be noticed by a denizen of the ocean floor. The solid Earth itself is also distorted but by a negligible amount since it is very rigid. But no substance is so rigid that it withstands any tidal squeezing. Consider the smaller Moon. A point on its surface is subject to 81 times the tidal force that we undergo because the Earth is 81 times as massive as the Moon. This is enough to force the all-solid Moon to show the same face to the Earth, and its shape has been warped such that its diameter along the radial direction is about a mile longer than its diameter in the transverse directions. Some of the satellites of the massive outer planets undergo even greater tidal stresses. The case of Io, the innermost of Jupiter's four large moons, is the most famous. It is subject to a Jovian tidal force some 25,000 times that undergone by the Earth. It orbits Jupiter once in only 42 hours in a slightly elliptical orbit, so this mighty tidal force is itself varying over that period. The rigid Moon-sized Io undergoes a regular squeezing like a tomato or peach being tested for ripeness. All of that extra energy heats Io's interior to the point that this small world is the most volcanic object in the solar system, all due to tidal forces.

The Moon has done far more to influence the Earth than to grace our night and provide fodder for songwriters. Its extensive alterations of our oceans, atmosphere, and habitat have been documented in detail. Probably the most impressive difference from a moonless Earth comes from its long-term effect on the tides. The tidal bulges of the oceans causes them to beat against the continents and the ocean floor and brake the rotation of the Earth, just as this planet has slowed the rotation of the Moon to equal its orbital period of $27\frac{1}{3}$ days. With no moon the Sun alone would not have had so great a tidal effect on the oceans, and the Earth would today rotate on its axis in only eight hours.

Imagine four hours of daylight alternating with four hours of darkness. The adjustment of circadian rhythms to this pace would have given rise to different life-forms. The great winds and storms caused by three times the present rotational velocity of the Earth (sometimes called the Coriolis effect) would be a trial for all land-based life-forms. These are but some of the many not-so-subtle differences that would be shown by a moonless twin of our world.

CHAPTER 14 Chiaroscuro

Every opaque body is surrounded and its whole surface enveloped in shadow and light.

—LEONARDO DA VINCI

Chiaroscuro, the treatment of light and shade, is a technique used in art to produce the illusion of depth and for a dramatic effect. It no doubt arose from observations of nature, and certainly of the night sky.

Leonardo da Vinci became its first great, and most famous, innovator. In his art and also in his science, he applied it to many things with success. It was he who first gave the correct reason for earthshine, the effect by which the whole Moon can be faintly seen during the crescent phase when its sunlit hemisphere is mostly turned away from the Earth.

Whatever phase of the Moon appears to us, we know that on the Moon, the Earth appears in just the opposite phase. As we see the thin bright crescent of the Moon, one standing on the Moon would see a gibbous Earth bordering on full, a very brilliant Earth about 4 times the diameter, and thus nearly 16 times the apparent area of the Moon as it appears to us. The percentage of incident sunlight reflected by the Earth, called the albedo, is several times that of the airless Moon. Altogether, the Earth outshines the Moon by at least 40 times, and this lights up the lunar dark side to the point where we can see it. The actual picture may be even more extreme. From eastern North America, for example, we observe the waxing crescent Moon in our western sky after sunset. At that time, the West Coast and the Pacific Ocean beyond are still in daylight, and a specular reflection of the sun off the Pacific waters can increase the brightness of the earthlight shining on the night side of the Moon. From Europe, the same role would be played by the Atlantic Ocean to its west. The fading of earthshine with the waxing of the Moon through the crescent phase into first quarter and beyond comes with the fading Earth, which appears—from the Moon—to slim down from the waning gibbous phase toward last quarter.

Leonardo, in some rather complex twists and turns of logical argument, came to this same conclusion. As noted earlier, most people associated earthshine, if at all, with a burnout of the Moon of the previous lunation. But Leonardo appears to have come up with a basically correct view, garbled somewhat by minor untruths along the way. In his notebooks, he states,

> The Moon has no light in itself; but so much of it as faces the Sun is illuminated, and of that illuminated portion we see so much as faces the Earth. And the Moon's night receives just as much light as is lent it by our waters as they reflect the image of the Sun, which is mirrored in all those waters which are on the side towards the Sun.... when the eye is in the east and sees the Moon in the west near to the setting Sun, it sees it with its shaded portion surrounded by luminous portions; and the lateral and upper portion of this light is derived from the Sun, and the lower portion from the ocean in the west, which sees the solar rays and reflects them on the lower waters of the Moon.... therefore it is not totally dark, and hence some have believed that the Moon must in parts have a light of its own besides that which is given it by the Sun; and this light is due, as has been said, to the above mentioned cause—that our seas are illuminated by the Sun.

Jean Paul Richter, compiler of Leonardo's notebooks, maintains that this passage establishes Leonardo's prior claim to be regarded as the original discoverer of the cause of the earthshine, about a century before Johannes Kepler and his teacher, Michael Maestlin, made the same claim independently (one of Leonardo's sketches is reproduced on page 141). In any event, Leonardo's

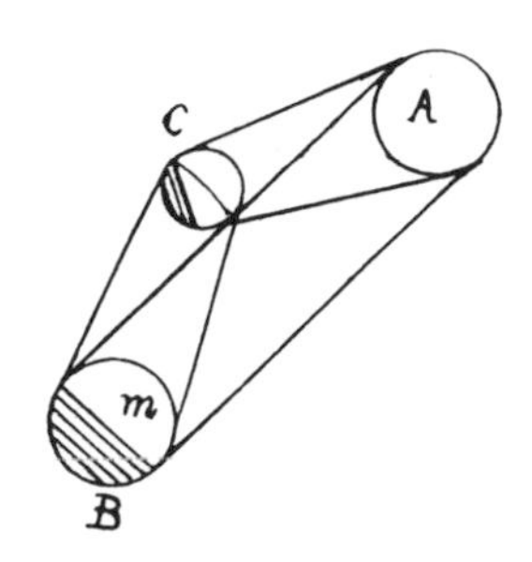

Drawing by Leonardo da Vinci allegedly illustrating earthshine and its correct explanation. A *represents the Sun,* B *the Earth, and* C *the Moon. From* The Literary Works of Leonardo da Vinci, *published in London, 1883.*

discourse reveals his belief that the Moon, like the Earth, was largely covered with oceans. Without a telescope, it would be natural to assume the two worlds to be similar in the main, a belief that if nothing else is a giant step from the pre-Copernican belief that the heavens are divine and made of the pure quintessence, unlike our impure world made of earth, air, fire, and water, as claimed by Aristotle.

* * *

How bright is the Moon? Obviously it depends on its phase. But how much does the phase affect the moonlight we see? The Moon when full shows twice as much illuminated area as it does at first or last quarter. But the full Moon is not just twice as bright as it is at the quarter but about ten times as luminous. The brightness of the Moon turns out to be highly dependent on the phase, which is determined by the phase angle, defined as the angle between the Earth and the Sun as seen from the Moon. This angle is about zero at full Moon and 90 degrees at either quarter. At the midgibbous and midcrescent phases (positions 2 and 4 in the figure on page 118), the phase angles are 45 degrees and 135 degrees, respectively, and moonlight is reduced to one-third that of full at the midgibbous position, and by as much as one-eightieth at midcrescent.

Those doubting such enormous changes from one phase to another, or even from one night to another, can measure them for themselves. They need only a sheet of white paper and an outdoor light source of constant luminosity. A nearby streetlight or yard light will do, providing that it does not change brightness with the direction (left, right, up, or down) from which it is viewed. As we approach and recede from it, it will brighten and dim as the square of our distance from it.

It is very difficult to determine the distance at which a light equals the Moon in intensity by looking at the two together directly, especially with a difference in their color. But if one holds a stick in one hand and a sheet of paper in the other in such a way that the shadows of the stick blocking both the Moon and the light fall on the paper, then by moving toward or away from the light, one can fix the two shadows (one from each source) to appear equally dark to a very precise degree. From night to night, the distance to the light yields a measure of the relative light intensity of the Moon.

Two phenomena account for the wide variation in lunar brightness with change of phase. The first is due to the effectiveness of incoming solar radiation. A terrace sloping toward the south is known to lose winter snow cover on its slopes more rapidly than on its level parts, because the sunlight is more nearly perpendicular to the slope and each square foot or meter of the surface receives more heat. The full and the quarter Moon are analogous to the slope and the level surface. At full, we view the Moon from the same direction as does the Sun, but at the quarter Moon the sunlight appears to us to come from the side, with each ray spread across a wider surface. Second, the Moon's surface is very rugged, and much of its terrain lies in shadow, as is easily confirmed by a view through a telescope. No shadows appear at full Moon, and consequently the telescope shows no relief and the moon appears brighter but less interesting as a result.

The full Moon shows twice the illuminated area as does the quarter phase and, being ten times as bright, must average to about five times the brightness per unit angular area. We are so used to seeing the Moon shining brightly up in the dark that we pass over its great brightness variation with phase. We do the same thing with regard to everyday scenes closer to home as well. The first figure on page 143 shows the first-quarter and full moons with equally bright surfaces, as they customarily appear

The Moon at the first quarter and full phases, commonly but incorrectly showing them as equally bright. Courtesy of Yerkes Observatory.

in most books on astronomy. But in truth the relative light intensities should appear as they do in the bottom figure on this page. In the set of photographs on page 144, the top photo was taken at midmorning and the bottom photo at noon. The right side of the house faces east and the left side faces south. The midmorning

The full Moon is correctly shown as about five times as bright as the first-quarter Moon for one hemisphere, or ten times overall. Courtesy of Yerkes Observatory.

The top photograph was taken at midmorning and the one on the bottom at noon. The right or east side of the house is brighter than the left or south side in midmorning since sunlight shines more directly on it. At noon the south side is brighter. Our eye tends to see the two sides as equal in light intensity, since we recognize that they reflect equally under identical lighting conditions.

sunlight strikes the east side directly and the south side obliquely, making the latter seem less bright to the eye and the camera. The situation is reversed at noon. We know the sides of the house to be equally reflective, and so our mind's eye tends to ignore the actual light difference. Artists learn to see these differences and use them; for example, Claude Monet made a series of paintings of the Rouen Cathedral under different aspects of light.

CHAPTER 15
The Planets

Seven celestial objects share a property borne by no other in the sky. They move, or rather appear to move, against the background of seemingly fixed stars. Despite their varied appearance to the eye, they have been identified as a group for many centuries. These are the planets. Not all of them are among the planets as we define them today (the Sun and the Moon are no longer considered among their number), but they do comprise the visible members of our solar system.

The Sun, the Moon, and the five true planets visible to the naked eye have together dominated our view of our cosmos. They gave their names to the days of the week and imbued the number seven with its special mystical significance. They also became the central players in the scientific revolution of the sixteenth and seventeenth centuries.

In any of the Romance languages, all seven days of the week can be clearly identified from their names. In English the days of the midweek, Tuesday through Friday, are named for Teutonic gods; however, they are roughly equivalent to, in order, Mars, Mercury, Jupiter, and Venus. Saturn, the Sun, and the Moon make up the balance of the week. Many are understandably puzzled over the order of the seven days, entirely scrambled as they are of any order of distance from either the Earth or the Sun. The scheme of the sequence derives from the fact that the day is divided into 24 hours, a very ancient practice. Prior to Copernicus and his heliocentric theory, the seven recognized "planets" were most commonly ordered from the Earth, with the Moon properly closest followed in order by Mercury, Venus, the Sun, Mars, Jupiter, and finally Saturn just this side of the sphere containing the stars. Back in those days, when astronomy was inseparable from astrology, one of the seven was believed to rule each of the 24 hours of the day. The order was reversed, with Saturn leading the list followed by the others in turn; thus the Moon ruled the seventh hour followed by Saturn, which ruled the eighth. Keeping to the inward-ordered scheme, we will find that Mars ruled the twenty-fourth hour and the Sun ruled the following hour, which began the following day. Thus Sunday, named for the Sun, follows Saturday. The continuation of this pattern produces an order of first-hour dominance that corresponds to the order of the days of the week, with each day taking its name from the ruling planet of its first hour. This is the explanation that has come down to us through the ages.

From their apparent motions in the sky, the magnificent seven can be divided into three groups. The first contains the Sun and the Moon. They are the two bodies that maintain about the same distances from us. The one orbits about us, and we circle about the other. Consequently, they march eastward across the heavens at a nearly constant rate. If we were to stop the Earth's rotation, by which all objects except the circumpolar stars rise and set, we could watch the Sun apparently moving about one degree, or twice its own apparent diameter, eastward each day. In about 365 days, it appears to move through 360 degrees along the entire Ecliptic. The Moon makes the same trip in only about $27\frac{1}{3}$ days and so moves about 13 times as fast as the Sun.

Since the true planets all circle about the Sun, they must sometimes be on our side of it and on the far side at other times. Their distances from us therefore vary by much greater ratios than those of the Sun, Moon, and stars. Furthermore, we are revolving on a moving platform in their midst. As a consequence, their motions as viewed from Earth are more complex than are those of the Sun and the Moon (which compose a "first group" of celestial objects with comparatively simple motions), and their speeds across the sky are far from constant. The motions of Mercury and Venus, which form the second group, resemble each other, whereas those within a third group that includes Mars, Jupiter, and Saturn move very differently. (Uranus, Neptune, and Pluto, worlds that can only be seen with optical aid and were not discovered until recent times, also belong to the third group.) The dissimilarity of Mercury and Venus from the other planets is due to their proximity to the Sun. Being always closer to the Sun than the Earth, they never appear to stray far from it. Astronomers call them the inferior planets, and the rest the superior planets, since they are all more distant from the Sun than is the Earth and can appear at times opposite the Sun in the sky.

Mercury and Venus appear to shuttle back and forth from one side of the Sun to the other; on one side they precede the Sun and are called morning stars, and on the other they follow it and set after it does. Then they are seen in the evening sky after sunset. In between times they are too close to the Sun to be seen. Mercury takes only four months to complete the journey from one evening appearance to the next; its proximity to the Sun assures that it flits about very rapidly. Small and nearly airless, Mercury still receives so much sunlight that it is one of the brightest things in the sky. Its proximity to the Sun limits its visibility in the twilight sky at dawn or dusk, hence it is only occasionally visible and never spectacular. Copernicus is said never to have seen Mercury, but if that is the case, he didn't try very hard. Even in the cloudy climate of northern Poland where he lived, this planet would occasionally be quite visible if not brilliant in the western sky, especially in late winter and spring.

Venus is altogether different. Only the Sun and the Moon outshine it, making it much the brightest pointlike object of all, and the only one easily visible in full daylight. Finding it in the daytime is not so much a problem of marginal visibility, but rather one of detection in the featureless blue sky. Once located it is not difficult to spot. At night, Venus's lambent glow is sufficient to cast shadows. Its orbit is not far within our own and it overtakes us only slowly. That is why Venus takes nineteen months, not four like Mercury, to pass through its configurations and return to the same relationship with the Earth and the Sun.

The superior planets seem to move entirely differently. Jupiter and Saturn take about 12 and 30 years, respectively, just to orbit the Sun once. The velocity of a planet in its orbit depends on its distance from the Sun, and these distant giants lumber along at less than half our orbital speed of about 20 miles, or 30 kilometers, every second. As they move ponderously along the ecliptic it can sometimes take days or weeks for us to perceive that they are moving at all.

Mars is not nearly so far beyond the Earth, and we overtake it less quickly than we do the outer planets. It takes less than two of our years for Mars to complete one full revolution about the Sun, and its velocity is just below ours. When Mars is opposite the Sun from Earth it is closest to us, and for a two-month period it appears to move westward among the stars, reflecting our greater speed as we pass it by. At all other times it moves in the normal fashion toward the east but not quite as fast as the Sun, since we are revolving about the Sun at a faster rate.

* * *

"I've noticed a very bright star over the last few nights in about the same place except that it moves when I watch it. Do you know what it is?" Many times over many years I have been asked questions like this by a student, friend, or stranger. A little probing almost always establishes that the motion is shared by all unearthly objects as they move from east to west across the sky (known as the sidereal motion), especially if the questioner has seen the star over more than one night. Although almost everyone realizes that the Sun and the Moon rise in the east and set in the west, not as many realize that the planets and stars do so as well.

I have responded to hundreds of such queries. This is no surprise in itself. The surprising element lies in the fact that the celestial object in question has always been one of three planets, Venus, Mars, and Jupiter, or one of the two brightest stars, Sirius and Canopus. The great majority of questions concern Venus, with Jupiter firmly in second place.

At first glance, it might appear unusual that so few objects elicit so much notice. But with one addition, this is just the group of starlike objects that shine with an apparent magnitude of −1 or brighter. Mercury can also appear of such brilliance, but this nearest of all planets to the Sun is only seen low in the twilight sky just before dawn or just after sunset and it often escapes notice for this reason.

If my experience is typical, I suspect that people may be perceptive to differences in the sky between the unusual and the commonplace. The reason for my suspicion lies in the great increase in numbers of starlike objects if the apparent magnitude limit of −1 (or −0.5 more precisely) is extended just one magnitude fainter to include those of zero magnitude (brighter than +0.5) as well. The list swells from 6 (including Mercury) to 15. It now includes Vega, Arcturus, Capella, Rigel, Procyon, Achernar, and Alpha Centauri, the last two stars being found in the deep southern sky. It also includes Betelgeuse and the planet Saturn most of the time. Both vary in light to the point where they can occasionally appear fainter than +0.5. Never are less than two among this group visible high in the sky at any time from any latitude, and usually more can be seen. Eight of the nine are bunched between magnitudes 0 and +0.5; only Alpha Centauri at −0.3 is brighter. Might even the casual observer, consciously or otherwise, be aware that some of the zero-magnitude objects are always around, whereas frequently no brighter object is visible?

The case of Mars warrants special consideration. Among the bright objects, Mars alone varies greatly in apparent brightness over time. At its very best it outshines every other object in the midnight sky but it is usually much fainter. In fact, most of the time it appears only at the first or second magnitude. Once about every two years and two months, Mars is lined up with the Earth and we see it opposite the Sun from us; this configuration is known as opposition. Around that time, Mars is best seen, like the full Moon, in the middle of the night. Opposition is the time of its closest approach to the Earth, and for a brief period surrounding it Mars brightens rapidly to become a very conspicuous object attracting widespread notice. Because its orbit is more eccentric than that of the Earth, its distance at closest approach varies considerably (see figure on page 152). The magnitude of Mars at these times of opposition and closest approach varies from about −1 to −3 and back to −1 over a period of about 17

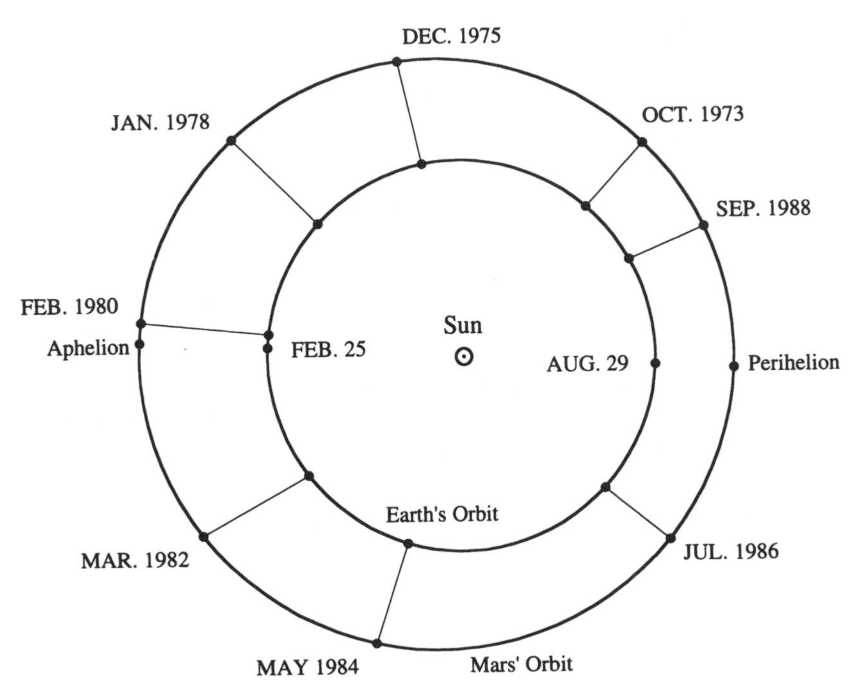

The orbits of the Earth and Mars to scale. The orbit of Mars is more eccentric and thus its distance from Earth at opposition varies. Mars can get relatively close, as it did in 1988, or far, as in 1980.

years. If we assume that −0.5 is a limit fainter than which public notice is minimal or absent altogether, we find Mars above this limit only 15 percent of the time over the last 17 years. During its most recent favorable opposition in 1988, it remained brighter than this limit for about 180 days and topped off at −2.8, just brighter than Jupiter at its brightest and second among starlike objects only to Venus, which shines between magnitudes −3.5 and −4.5. At its unfavorable opposition in 1980, Mars reached only −1.0 and remained brighter than −0.5 for only 56 days. Since Jupiter always appears within the magnitude range extending between about −1.5 and −2.5, it is almost always brighter than Mars and is understandably noticed much more frequently and regularly.

Everyone knows that the Moon can often be seen in the daytime. But it is not commonly realized that under ideal conditions, other objects can as well. Venus is the brightest of the starlike objects in the sky; it is outshone only by the Sun and the Moon. As noted earlier, it can cast shadows at night, and it is visible in the daytime sky whenever it is displaced well away from the Sun and the sky is clear.

Under rare conditions, another planet and at least one star are also visible to the naked eye. I have seen Jupiter several times just after sunrise or just before sunset, and once even the brightest star, Sirius, as well. Since a clear blue sky may only be one-tenth to one-thirtieth as bright when the Sun is at the horizon as it is at noon, marginally visible stars and planets are much more visible at that time. In the case of Sirius, I had not only that advantage, but I also happened to be at the Kitt Peak National Observatory in southern Arizona, where the 7000-foot altitude reduced the amount of air through which Sirius must shine, and I was just able to see it as the Sun set.

Theoretically, Mars should also be seen in the daytime since at its very brightest it outshines even Jupiter. But it does so only when it is opposite the Sun in the sky and then it is always near the horizon or below it in the daytime. It can never be seen at the optimum time as can Sirius and Jupiter. One other object potentially visible at its brightest is Mercury, but being as close to the Sun as it is, it is never likely to be seen in the daytime. All of these sightings presume average eyesight, and there are inevitably a few individuals whose eyesight is extraordinarily keen. We cannot rule out the possibility that someone, sometime, has spied Vega or Arcturus or Canopus just before sundown.

* * *

The Earth is one of four planets in the inner part of the solar system, the region closest to the Sun that with Mercury, Venus, and Mars forms the group of terrestrial planets, so called because they share a number of common physical properties that are not found elsewhere.

The Moon can be considered as a fifth member of the group, a fifth planet in a sense, because it is large for a satellite and because it shares the characteristics that identify the terrestrial group. The Moon is one of the seven large satellites in the solar system, all larger than Pluto, the smallest planet. If the Moon orbited the Sun directly, instead of the Earth, it would properly be considered a planet.

The four terrestrial planets and the Moon are of reasonably similar size and mass when compared to the four major, or giant, planets—Jupiter, Saturn, Uranus, and Neptune—and each consists of a solid body with a dense, largely iron core encased in a solid mantle, which is in turn surrounded by an atmosphere of relatively small vertical extension, or almost none at all in the cases of the Moon and Mercury. They are the densest bodies in the solar system and as mentioned earlier are the closest to the Sun.

It is well known that these five worlds were formed over four and a half billion years ago from the same primordial interstellar material from which the Sun and the major planets originated, mostly hydrogen and to a lesser degree helium. More than 97 percent of all material in the universe consists of these two lightest and simplest of all elements. The Earth and its terrestrial neighbors ended up, however, being almost totally composed of heavier elements such as carbon, oxygen, silicon, aluminum, and iron. The reason for this chemical anomaly lies in their relative proximity to the Sun. At the time of their formation the heat from the Sun and the tidal shear resulting from its powerful gravitation drove off most of the two lightest elements, leaving behind the small dense cores made of heavier stuff. Unlike the major planets that were able to retain most of their original material and their thick, dense atmospheres dominated by hydrogen, the planets of the inner solar system condensed into small, airless worlds. Later the seismic and volcanic activity within them spewed out other gases, in a process called outgassing, that allowed secondary atmospheres to form.

It is natural to expect that internally the earthlike planets would resemble each other, and indeed there are many more similarities in their constitution than those described here. Yet, the composition of their present atmospheres also reveals many differences between them. It is not too surprising that the Moon and Mercury have scarcely any atmospheres at all. Mercury is the closest planet to the Sun, and its high surface temperatures combined with its relatively low mass have not permitted it to retain more than a trace of an atmosphere. The Moon is virtually airless for the same reason; although it is farther from the Sun than Mercury, its mass is smaller. If the Moon and Mercury were much farther from the Sun than they are, they would be colder and able to retain a number of gases in their atmospheres.

The importance of atmosphere and clouds in the *albedo* (a term meaning the percentage of total incident sunlight reflected directly back into space) of a planet can most easily be appreciated by an examination of the atmospheres of the three remaining planets of the inner solar system. Venus, entirely enshrouded by clouds, has a very thick atmosphere. Mars, in contrast, has a thin atmosphere in which clouds are not commonly found. The albedo of Venus is over 70 percent, whereas that of Mars is only 15 percent. The Earth is in between, being partly covered with clouds, with an albedo near 35 percent. Mercury and the Moon, with scarcely any air at all and certainly no clouds, have albedos of only 7 percent. The ordering of these inner planets (and the Moon) with declining thickness and cloudiness of atmosphere from Venus to Earth to Mars to Mercury to the Moon is one of declining albedo as well.

The three largest planets of the terrestrial group all have substantial atmospheres, but their surface air pressures are very different. The pressure on Venus is about 90 times that of the Earth. In fact, its thick mantle of air is about as dense as ours would be if all of our oceans boiled off into steam. Mars, on the other hand, has a pressure only about one-hundredth of ours. Carbon dioxide constitutes about 95 percent of the atmospheres of both Venus and Mars, whereas it accounts for only a few hundredths of one percent of the atmosphere of the Earth. Noteworthy also is the extent of free oxygen in the Earth's atmosphere (21 percent), where it is the second most abundant element after nitrogen (78 percent); both of these gases are nearly absent in the atmospheres of the other two planets.

We know that originally the Earth, too, had an atmosphere dominated by carbon dioxide. But quite early in our planet's history, the self-replicating stuff called life formed. Although simple at first, one-celled bacteria and algae could use solar energy to break down carbon dioxide and water and to release oxygen in a process known as photosynthesis. Plants still use this process, which enriches our air with oxygen.

Our planet is thus unique, at least among the objects we know, by being mostly covered by oceans, having abundant free oxygen in its atmosphere, and as far as we know, being the only source of life. It is indeed the pearl of the solar system.

CHAPTER 16 The Outer Limits

The great tragedy of science—the slaying of a beautiful hypothesis by an ugly fact.

—THOMAS HENRY HUXLEY

Three distinct planets have not been known throughout history as have the brighter ones. These are Uranus, Neptune, and Pluto. There has long been speculation that the list of planets is not yet complete, and that one or more still remains undiscovered in the outer reaches of the solar system. To place this theory in perspective, we must examine these three distant worlds and the nature of their remote domain.

Uranus was discovered in 1781 by Sir William Herschel, a musician turned astronomer living in Bath, England. His telescope was of sufficient size to see the planet as a greenish disk. He first thought he had discovered a comet, but its slow orbital motion revealed it to be far too distant. He then recognized it as a planet, the first discovered since prehistoric times. He first named it Georgium Sidus after his monarch, King George III, but collective wisdom prevailed and it was named Uranus after the Greco-Roman god of the sky.

Under the best of circumstances, Uranus can just be seen with the naked eye. How, then, did it remain undiscovered until so late? The reason is that although visible and seen many times before 1781, it is close to the threshold of human vision, along with many stars. Mapping them, and noting that this one moved among them, was just not possible.

The discovery of Neptune in 1846 was not serendipitous; it was a triumph of Newtonian mechanics. A popular conception of discovery finds expression in the lines of John Keats:

> Then felt I like some watcher of the skies
> When a new planet swims into his ken.

But Keats only knew of the discovery of Uranus, and the first four asteroids in the years 1801–1807, all of which were found by chance, and not from the use of physical laws. In the years following the identification of Uranus as a planet, its orbit was followed closely and found not to proceed along the orbit prescribed for it by calculations using Newton's laws and subsequent developments in celestial mechanics. The discrepancies were small at first, but soon became disturbingly and irreconcilably large. The idea dawned slowly that another, unknown object was perturbing the orbit. Two astronomers (J. C. Adams in England and U. J. J. Le Verrier in France) made the laborious calculations to locate the new planet in the sky, and when told where to point his telescope, J. G. Galle of the Berlin Observatory looked and discovered Neptune, named after the Roman god of the sea, in 1846. Neptune can be seen on a good night with a good pair of binoculars. But, again, the trick is to identify which of many faint points of light is the planet and not a star. Galileo had seen it over two centuries earlier and considered it to be a star; had he seen its motion he might have recognized it for what it is.

All appeared right with the Newtonian cosmos until later observations revealed over time that these two outer planets still strayed from their preordained orbits. Once again an unknown mass had to exist somewhere out there, pulling its neighbors away from their proper tracks. At his observatory near Flagstaff, Arizona, Percival Lowell had set about the task of finding the missing planet in the 1920s. While Lowell did not live to see it, in March 1930, Clyde Tombaugh, a young astronomer on the Lowell staff, identified a very faint speck by its motion to be a planet. It was named Pluto, partly because the first two letters formed the initials of Percival Lowell.

The new planet received much popular attention, even lending its name to Mickey Mouse's dog. As the faintest and most distant thing in the solar system, Pluto and its orbit were then considered to be its outer edge. Gustav Holst, the English composer who some years earlier had written his masterpiece, the immediately popular orchestral suite known as "The Planets," lived for years after Pluto was discovered. To be sure, the extramusical impetus for the work came from his rising interest in astrology, and the mystical Neptune forms a fitting concluding movement. But whether Holst saw the planets as gods or as worlds, he must have been asked if he might be induced to add a movement for this newfound planet.

* * *

Pluto is faint, so faint that it takes a substantial telescope to see it, and it does not appear as a disk to the eye in even the largest of telescopes. For this reason attempts to measure its size were foiled for many years. To account for the discrepancies in the orbits of the other two outer planets, Pluto had to be more massive than the Earth. But its failure to be seen as a disk in the largest telescopes placed an upper limit to its size near that of Mars, about half the Earth's diameter. Pluto would have had to be by far the densest object known, far denser than the densest-known planets: Mercury, Venus, and the Earth, each about five times as dense as water. No abundant substance is present in the solar system from which such a dense planet could be formed.

For years, every time larger telescopes and more sensitive detectors were used to make a better derivation of its size, Pluto came out smaller and smaller again. Pluto is a mystery—in many ways. Its orbit is at once more eccentric, more of an elongated ellipse, and more highly inclined to the plane of the solar system formed by the rest of the planets' orbits than that of any other planet. Its path is so irregular that at its closest approach to the Sun—where it has been since 1980 and will be until the year 2000—it comes closer than Neptune. However, there is no danger of collision between the two outer worlds because of the different inclinations of their orbits. Pluto moves slowly and takes almost 250 years to circle the Sun once, whereas Neptune takes

165 years. Neither planet has made even one full trip around the Sun since its discovery, but Neptune has not far to go; it will reach its point of discovery in the sky by about 2010. Pluto, however, will not do the same until 2180, almost two centuries later.

The reasonably close encounters between Pluto and Neptune have led some astronomers to the theory that Pluto was once a satellite of the larger world. Neptune has one large moon, Triton, with a mass big enough to have made such a triad unstable, so one of the two smaller bodies would have been quickly ejected (quickly on an astronomical time scale) from the system. Triton moves about Neptune in a retrograde orbit, that is, in a direction counter to that in which all planets go about the Sun, in which most of them rotate, and in which most moons circle their primary planets. Triton could have reversed direction in the brouhaha. Although we may never know if such a disruption actually occurred, it is plausible that it may have happened not long after their formation about 4.6 billion years ago.

In 1979, Pluto was found to have a moon. Once the period of its revolution about the planet and its mean distance from the planet were determined, the mass of Pluto became well known from a straightforward application of Newton's laws. Since we now know that things in the outer solar system except the four major planets are mostly composed of water ice, we know Pluto's diameter as well. Few astronomers were surprised to discover that this outermost planet turned out to be smaller and much less massive than our own Moon. No fewer than seven satellites are known to be larger than Pluto; in addition to the Moon, there are the four big moons of Jupiter (Io, Europa, Ganymede, and Callisto) first seen by Galileo in 1610, Titan, the largest satellite of Saturn (discovered by C. Huygens in 1655), and Triton (found by N. Lassell in 1846, just weeks after the discovery of Neptune itself). One astronomer has remarked that if it had been discovered today, Pluto would have been called an asteroid (when the remark was noted by the media, Pluto's discoverer, Clyde Tombaugh, let it be known that he did not agree!).

The distinction between planet and asteroid is more one of semantics than size or mass. Pluto, all other planets, and all seven of the larger moons possess atmospheres, though remarkably thin in some cases. But in Pluto's case, the matter of gravitational attraction was clear; by no means could this small ice world disturb significantly the orbits of the two giants Uranus and Neptune. It was concluded, then, that yet another unknown world was out there. One more body, Earth sized or larger, was thought to wheel about the Sun in an immense orbit beyond Pluto. Since it would be a tenth planet, it was known as planet X.

Matters stood at this point until 1992, when the masses and orbits of the outer planets had been more precisely measured by the *Voyager 2* space probe that passed by Uranus in 1986 and Neptune in 1989. After a detailed analysis, *Voyager*'s passage required very small mass and orbit corrections to both—and the need for perturbations went away! No longer does planet X need to be summoned to fix the Newtonian mechanical system; in fact, a large mass out there is very unlikely. More Plutos may well be found, including one in a highly inclined orbit twice Pluto's distance with a period near 800 years, but it would only be a few hundred miles in diameter. Nothing of much greater bulk is likely to be waiting for us to discover and probe.

In our roundup of the solar system, we must turn to the innermost planet, Mercury. Mercury's orbit is almost as eccentric as Pluto's, and it orbits the Sun faster than Newtonian mechanics allows. Something had been known to be wrong for more than a century, but in this case, the palliative in the form of an undetected planet didn't work. A planet close to the Sun, interior to Mercury, was sought, and purportedly seen by some observers. It was even given a name, Vulcan, after the Roman god of fire and the hearth. We know now that Vulcan does not exist. Early in this century, the one accepted alteration or generalization of Newton's universe was shown to explain Mercury's motion. It took Albert Einstein to accomplish this feat.

Jacob Bronowski stated, in *The Common Sense of Science,*

> No law ever gave wider satisfaction than the law of gravitation. Yet we have seen that the explanation it gave of the workings of nature was false, and the understanding we got from it mistaken. What it really did, and did superbly, was to predict the movements of the heavenly bodies to an excellent approximation.

In his special and general theories of relativity of 1905 and 1915, Einstein identified matter and energy as interchangeable states in a single concept when he equated them in the relation given by $E = mc^2$.

Newton's shortfall derives from his picture of time, which has no direction at all, and could run backward as easily as forward. His laws are approximations and there came a point when the approximations broke down. When that point was reached in astronomy, Einstein's laws took their place, and later precisely accounted for the discrepancy in Mercury's orbital motion. In Einstein's cosmos gravity attracts and bends light, and this, though seemingly absurd on the face of it, has been proven over and over again. Even now, the Hubble and other space telescopes and probes are discovering other effects of the gravitational bending of light, as Einstein's theories predicted. Einstein is the most famous scientist of the past three centuries for good reason.

Neptune, Vulcan, and Pluto are the three objects in the solar system that were postulated before discovery. All caused disruptions in the grand Newtonian scheme of things, but one, Neptune, went on to be discovered to be a major planet, thus righting the framework of classical celestial mechanics, while the second, Vulcan, was never found. If it exists at all, it would be much too small to be a planet or even a sizeable asteroid, certainly nothing that could shake Mercury from its path. Pluto fell somewhere in between, at first seeming to confirm but later being found too small for the role intended for it. What caused Neptune to preserve and Vulcan to help to depose the reigning Newtonian paradigm? Observations surely, spotting Neptune almost at once, while searching for Vulcan in vain, and a century later finding the raison d'etre for a substantial Pluto to vanish. But the observations were motivated by theories, right and wrong. Taken together the three illustrate just how science is done.

CHAPTER 17
Calamity!

I am become death, the shatterer of worlds.

—BHAGAVAD GITA
Spoken by J. R. Oppenheimer upon the detonation of the first atomic bomb at Trinity, New Mexico

How old are the craters on the Moon? We know that from about 4.6 billion years ago, when the Sun and planets were formed, until around 3.8 billion years ago, the young solar system was a tough neighborhood. Much of the original debris remained on the loose even after the planets and larger moons and asteroids had condensed out of the primordial ooze. The smaller chunks were many many times as numerous as they are today. But through collisions with the large bodies and with each other, they thinned out, and the last few billion years have been much quieter and less prone to catastrophic events. Most of the Moon's surface as we see it today was already in place two or three billion years in the past.

We now know that the last giant asteroid impact on the Earth occurred about 65 million years ago. Along about 1980, Luis and Walter Alvarez, father and son, physicist and geologist, along with a few colleagues, theorized that the dinosaurs were done in at that time, not by a slow succession of alterations in the geologic and climatic conditions of the day but by a very sudden catastrophic occurrence. They speculated that a large comet, or more likely a large asteroid, collided with the Earth. This has become one of the most widely known and popular theories of the century, and has more or less resolved the long-standing persistent debate between catastrophism and uniformitarianism. These are terms for the doctrines that hold that nature changes relatively suddenly and massively, and alternatively, that geologic and other changes come about only very slowly through gradual evolutionary processes.

Throughout the late nineteenth and early twentieth centuries uniformitarianism held sway, but more recent evidence for short and calamitous events has grown, and in many cases prevailed. The Alvarez theory was disdained by many scientists, partly because of their residual belief in a measured pace as the only medium of change. But since its introduction, its acceptance has grown, and in 1991, the long-sought site of the impact was discovered. This smoking gun, in the form of a huge, vestigial crater, known by the Mayan name of *Chicxulub* and measuring well over 100 miles in diameter, was found and dated to the time of the disaster, 65 million years ago.

It is today widely accepted that a sizable asteroid of perhaps five to ten miles in diameter on a northerly course slammed into the coast of what is today the Yucatan Peninsula and the Gulf of Mexico at a relative speed of 20 to 40 miles per *second.* This greatest cataclysm of the past 200 million years or more threw aloft cubic miles of dirt and dust into the stratosphere, far more than would rise from a total nuclear exchange emptying all of our atomic arsenals. There it remained for months. The entire globe was plunged into a cold winter as all but one percent of the available sunshine was blocked by a cloud-enshrouded atmosphere like the present atmosphere of Venus. Events of this magnitude are sometimes referred to as mass extinctions, because so many species pass into oblivion almost at once. This impact is known as the K/T event, since it defines the boundary in time between the Mesozoic and Cenozoic eras, and more specifically, the Cretaceous and the Tertiary periods (the *K* in K/T derives from the German spelling of Cretaceous, which begins with the letter *K*).

During the 3000 years that elapsed before stable conditions were restored, more than one-half of all species of plants and animals vanished, including all remaining dinosaurs, the great saurians that had ruled the land for 140 million years. Their disappearance may have allowed the tiny mouse-sized mammalia to grow and eventually take over the remade world.

Could such a disaster happen again? Yes, of course it could—there are still large chunks of stuff circling round the sun. The acceptance of the Alvarez theory led to a remarkable increase in interest in asteroids and comets, the oft-ignored debris of the solar system, left over from its creation. The hit movie *Jurassic Park* and some other media events have embellished at least one supposed consequence of that great event and the fascinating theory behind it. Had the dinosaurs lived, they speculate, might some of them have evolved into beings with an intelligence on a par with our own, and if so, might they have prevented the rise of *Homo sapiens*? And, therefore, is it not plausible that intelligent beings living elsewhere in the galaxy, if any, may be reptilian in form and function, and not mammalian?

Fortunately for our species, not many asteroids of this size are still around, and most of those that are have well determined orbits. Few icy comets are of this large size, although the recent spectacular Comet Hale-Bopp is indeed of the required mass to finish off the lot of us. Smaller mile-sized celestial detritus able to take out a continent at one shot are yet plentiful. Some day, with nuclear missiles at the ready, we may manage to defend ourselves. A nuclear detonation alongside an errant comet or asteroid could blow it into slightly different orbit, thus causing it to miss the Earth. Other tricks to accomplish the same result less destructively may also turn out to be feasible.

As a result of this perceived threat of a collision, the search for all significant objects in orbit around the Sun has gone through a resurrection. The largest problem in the scenario of a deterrent, nuclear or otherwise, lies in the shortness of the time period over which we would know that another K/T-type event was headed our way. If a similar-sized asteroid were aimed at the Earth today, the period of naked-eye visibility would be mercilessly brief. From the time that the intruder appeared as large as the full Moon until impact might be as little as a minute or two in the daytime sky; at night the object would first appear as a bright star perhaps an hour before impact. To be sure, we have other means of detection at our disposal, but the warning time would be most increased if we were aware of the orbits of all of the potential "earth-crossing" asteroids, meaning those few that come as close to the Sun as the Earth's orbit. A complete census of at least the very large asteroids will help to prepare us, but when all is said and done, the insurance of extreme infrequency is almost all that is left to us.

Along with asteroids, comets also have the potential to change the course of life through mass extinctions. Asteroids, or minor planets, are the small, mostly rocky bodies that circle the Sun directly. Most are found about midway between the orbits of Mars and Jupiter; collectively they are known as the asteroid belt. The largest, Ceres, is about 600 miles in diameter, about twice the size of the next two largest. About 30 are larger than 100 kilometers, or 60 miles, in diameter, and 300 are larger than 10 kilometers, or 6 miles, the size of the despoiler of the dinosaur's world. Most of these are in their proper place in the asteroid belt, and it is very unlikely that any would get gravitationally perturbed into an orbit in contact with ours. Continued observation assures that any future problem of this magnitude will be well known in advance. When smaller asteroids are considered, the problem worsens. About 6000 asteroids of about one-half to one mile across are known and named. But there may be as many as ten times this number out there.

Comets are part of the icy solar system. Along with Pluto and most of the satellites of the outer major planets, they are mostly composed of water ice. "Dirty snowballs" is the well-known description of them because tiny rocky and dusty particles are often embedded in the ice. As a comet moves into the inner solar system (roughly the asteroid belt and closer) the greater influence of solar radiation heats and activates them to the point where the tiny dust particles are driven away from the Sun. This forms a great tail, the spectacular part of a comet's show that can appear straight or curved and that is directed mostly away from the Sun, regardless of the direction of the comet's own orbital motion. A second type of tail, usually fainter and more bluish in color, is formed of ionized gaseous material that points more directly away from the Sun. Comet Hale-Bopp exhibited both kinds of tails at its brightest, but the blue ion tail was rarely visible except to the fortunate few who were well away from any light pollution.

The comet factory lies beyond Pluto, out to as much as 10 to 50 thousand astronomical units from the Sun in the Oort cloud, so called after its discoverer, J. H. Oort. It consists of millions of mostly small, dirty snowballs, and a few larger ones. They hang around out in space until some external force, a collision or gravitational perturbation, throws them into orbits directed inward toward the Sun and planets. Recently a somewhat controversial hypothesis has claimed that small snowballs, perhaps the size of a small house, collide with the Earth's upper atmosphere and burn up in it at the rate of one every few seconds. Observations have yet to fully confirm this idea, but if it is so, these snowballs, too small to do any damage, may collectively add to the water in our oceans. If confirmed, these ice particles are very likely to originate in the outer reaches, and probably the Oort cloud itself, where water remains in solid form until it enters the inner solar system where we live.

The most recent collision of any magnitude in our neighborhood may have taken place not on the Earth, but on the Moon. In Canterbury, England, in the year 1176, just six years after Thomas Becket was murdered by four knights in Canterbury cathedral, six monks all reported seeing the crescent Moon one evening swell into an irregular shape. Not all reports of this kind are entirely credible, then or now, but taken at face value, the monks most likely witnessed a large meteor or small asteroid strike the Moon near the edge of its visible side. A very young crater has been found near the probable spot, but "young" on the lunar surface could mean anytime between 800 and 800 million years ago. Not until we visit the site can the age of the young crater be more closely determined.

The Earth remains relatively unscarred compared to the Moon. Our world, being much the larger, has been much more bombarded than the Moon, but it is geologically and seismically active, and therefore has the means to resurface itself and cover the countless craters of the past. Two-thirds of the surface is covered by ocean, and plate tectonics create a continual process of continental drift, with overlapping and subduction ever occurring at the borders between the plates. Venus and, to a lesser extent, Mars are seismically active or have other means for obliterating old craters, and the four giant planets are enshrouded in thick atmospheres. But except for these and, of course, the Sun, all other objects in the solar system remain littered with craters from the early bombardment, beginning with the formation of the solar system 4.6 billion years ago and declining greatly about 3.8 billion years ago. Even the two tiny Martian moons, far and away too small for geologic and volcanic activity at any time, are heavily cratered, thus confirming the meteoritic origin of most craters everywhere.

Thus only very recent strikes remain to be seen on the Earth. Two events mark the greatest of what very recent history has to offer. About 25,000 to 50,000 years ago, northern Arizona was struck by a meteoroid or small asteroid of high iron and nickel content. It may have been about 150 feet, or 50 meters, across, and weighed as much as a million tons. It moved 10 to 15 miles per second just before it struck the Earth, hurtling from the stratosphere to the ground in its last second of life. Most of it shattered or vaporized upon impact, but not before it gouged out a crater in the crust about 4000 feet in diameter and 600 feet deep at the center. Many pieces of it have been found, some weighing as much as a ton, but no larger pieces of it exist. It is the high rate of speed that gives a projectile like this its great potential for disaster.

This great hole in the Earth, known as the Barringer Crater, is located not far from Winslow, Arizona, and it remains a major tourist site. It is probably the best known and most visited crater of its kind anywhere in the world. Shaped rather like a stadium set largely below the surface of the surrounding desert, it is of a far larger scale than any amphitheater men have built. If lined with seats for a sporting event, it could comfortably hold some six million spectators, more than the entire population of Arizona! Despite its great potential capacity, it would pose problems as an amphitheater—and not only because of its steep sides. Most people would need binoculars to see anything of the event being staged at its center. Those in the rear seats would do well to see the ring in a prize fight, much less the two contestants. And the audience of a ballet would hear the music delayed two full seconds after the actions of the dancers.

In recorded history, our most devastating celestial visitor was probably a small comet that smashed into remote northern Siberia on June 30, 1908. No human casualties are known from this occurrence, called the Tunguska event, although trees were felled for tens of miles in all directions. The comet hit in the daytime, yet reindeer herders hundreds of miles away saw the glare and heard the noise of the explosion. Our most spectacular and oft-witnessed event involved no collision at all. On August 10, 1972, a huge, glowing fireball, made by a meteoroid as big as a box car and weighing tens of tons, passed over the American and Canadian west in the daytime. Many vacationers noticed it as it moved over Utah up through Alberta, passing over Grand Teton, Yellowstone, and Glacier National Parks along the way. After skimming along in our upper atmosphere nearly parallel to the ground, it proceeded back out into space, not much the worse for wear. Had it landed it would have made a large crater suitable for a tourist site. Somewhere out there this chunk orbits the Sun, somewhat smaller now and following a different path from the one it had before the encounter.

Smaller meteor falls happen every year. They typically leave fragments a few pounds in weight and do little damage. By definition, a chunk that has reached the ground is a meteorite, while the object in orbit before the fall is known as a meteoroid. The boundary between large meteoroids and small asteroids is not defined. Meteors are those "shooting stars" that are too small to reach the ground, burning up in the atmosphere as friction heats them to thousands of degrees. Most are no bigger than a grain of sand.

Collisions with objects the size of the one causing the K/T event are rare; perhaps one per hundred million years is the average rate. Smaller ones a mile in size may come in with a frequency of one in a million years. The odds are still heavily on our side.

CHAPTER 18
The Last Return

Halley's Comet is neither the brightest of comets nor the most frequently seen, but it is the only bright comet that returns within a relatively short period of time. Most comets have retained their original orbital periods lasting many thousands of years; thus each of the bright ones has graced the inner solar system with its spectacular tail no more than once in recorded history. The orbits of a few, however, have been affected by one or another of the four giant planets. Jupiter, Saturn, Uranus, and Neptune each have sufficient gravitational pull to perturb the orbit of a comet into a period of only a few years or decades. This brings about more frequent periodic visits to the region of the inner solar system near the Sun and the Earth. Most comets of either orbital group are faint—too faint to be visible to the naked eye, much less spectacular, at any time. Of all comets, Halley's alone has been seen and recorded as a spectacular object more than once.

Since its orbit was perturbed, probably following a close approach to Jupiter a few hundred thousand years ago, it completes an orbital swing that brings it near the Earth every 75 to 78 years. Since 240 B.C. at least, its appearance has been recorded at every one of its close approaches, now 30 in all. Its great fame rests on the frequency and majesty of these visits and its repeated influence over human affairs. That majesty has taken many forms and enriched our individual and collective heritage in many ways, but its end may now be in sight.

Consider the Earth and its inhabitants from the comet's point of view. For the last few millennia it has been treated to a series of regularly spaced tableaux of human life, a sequence of stills illustrating our travails in detail but briefly. Thus, for instance, it bore witness in A.D. 451 to the Battle of Châlons, at which Attila the Hun was defeated by the Roman general Aetius.

Its most renowned visit came later, in 1066, when the comet appeared on the eve of the Battle of Hastings and (it is said after the fact) was taken by William of Normandy to be a good omen and by King Harold to be a bad one. For its part in that event its likeness was woven into the Bayeux Tapestry, chronicling the events surrounding the Norman invasion. Just after its unusually bright appearance three visits later in 1301, the Florentine master Giotto depicted it in "The Adoration of the Magi," one of his best-known paintings.

Five trips later in 1682, its fame was assured for all time. It was seen in that year by Isaac Newton and his younger colleague Edmund Halley. Prior to that time, comets had been thought to be transient phenomena in our atmosphere, not unlike rainbows, and as such they were taken to be auguries of momentous good or evil events to come. But Halley, fresh from learning of Newton's monumental work on the laws of motion and gravitation, perceived that the bright comets of 1531, 1607, and 1682 followed very similar paths and therefore must be the same comet with an orbit around the Sun that was readily calculable from the older man's theories. Halley made a prediction that the comet would burst into our skies again in 1758, some 75 years in the future.

Halley's audacious prediction was not forgotten and on Christmas Day in the year 1758, the comet was discovered right about where he said it should appear. Its verification of Newton's laws astounded and deeply affected the educated on both sides of the Atlantic Ocean as the full realization of their sweep sunk in. A middle-aged printer in Philadelphia and a young surveyor and a teenager, both of Virginia, all attested to Newton's impact on their thinking, how his laws showed them that natural laws governed the orderly cosmos. Later, Benjamin Franklin, George Washington, and Thomas Jefferson would put other unalienable laws to use for the government of humankind.

The comet's following two visits in 1835 and 1910 each occurred within weeks of the birth and death of America's foremost humorist, Mark Twain, who lost no time in the last weeks of his life calling attention to this coincidence. The latter appearance was not a spectacular one, because the bright part of the cometary visit occurred when it and the Earth were on opposite sides of the Sun. In fact, just months earlier in the same year, another comet appeared that had outshone Halley's. But Halley's is the comet remembered for its impact on the lives of the famous and the not so famous alike. Mark Twain is quoted as remarking: "I came in with Halley's Comet in 1835. It is coming again next year (1910) and I expect to go out with it. It will be the greatest disappointment of my life if I don't go out with Halley's Comet. The almighty has said, no doubt: 'now here are these two unaccountable freaks; they came in together, they must go out together.'"

In the year of Twain's demise, 1910, a boy of 13 went outdoors in a small town in northwestern Wisconsin to see each comet as it swung by the Earth. As my father was to tell his friends many years later, he did not confuse the one comet with the other. "I saw the first comet in January and Halley's not until late spring," he said, "and anyone can tell the difference between winter and mosquitoes in northern Wisconsin. Halley's was a disappointment, being neither very bright nor conspicuous." Nonetheless, he remained impressed with the sight of the famous object.

In the final year of my father's life the famous celestial visitor returned. He saw it once again, on his eighty-ninth birthday in January 1986, this time from the warm southeast coast of Florida where he lived in retirement. It was nicely visible with binoculars and just detectable without optical aid during this poorest of well recorded apparitions. Thus he became one of a select group of people, a "second timer," one who had seen Halley's Comet on two separate visits and was old enough to remember the first while seeing the second.

The visit was a disappointment, poor even when compared to its previous appearance. Nonetheless, no other object marks lifetimes as does this most famous comet. At its appearance in

1910, the crowned heads of Europe were massing in their horse-drawn carriages at the funeral of King Edward VII at the end of his gilded age, and in 1986 it witnessed the establishment of a national holiday to honor a martyred black minister. The world's population had tripled in the interim, bringing with it new problems such as global warming and AIDS, that my father could not have imagined when he saw it for the first time.

By the time the comet was visible in 1986, astronomers had developed and attached to their telescopes sophisticated new detectors that utilized a technology unimaginable only a few years earlier. These sensitive new detectors, called charge-coupled devices, or CCDs, permit the observation of the comet around its entire orbit, even when it is at its most distant point beyond Neptune near the edge of the solar system. It is being monitored regularly even now as it races outward past the giant planets. As so often happens, this gain in technology is offset by the loss of the excitement that had been felt in anticipation of the comet's return. For the first time since Halley made his daring prediction, the exhilaration of discovery is gone, along with the wonder at the reminder that heaven's immutable laws still hold. Even before the great comet flares up again in 2061 and beyond, the select few astronomers with access to these instruments can be second-, third-, and many-timers as they wish, transforming the comet from interloper into just another member of the solar family.

* * *

The 1990s have done very well by comets. Three among them have entertained and educated us, each in its own way. The first was Comet Shoemaker–Levy 9, named after its three discoverers, the comet-seeking couple Eugene and Carolyn Shoemaker and the astronomer-author David Levy, who first saw it on March 25, 1993.

This comet was very literally "captured" by Jupiter sometime before 1970 when that big planet so perturbed its path about the Sun that it was drawn into an orbit about Jupiter itself. Later, before its discovery, the comet passed within 21,000 kilometers, or about 13,000 miles, of Jupiter's surface. In the process, Shoe-

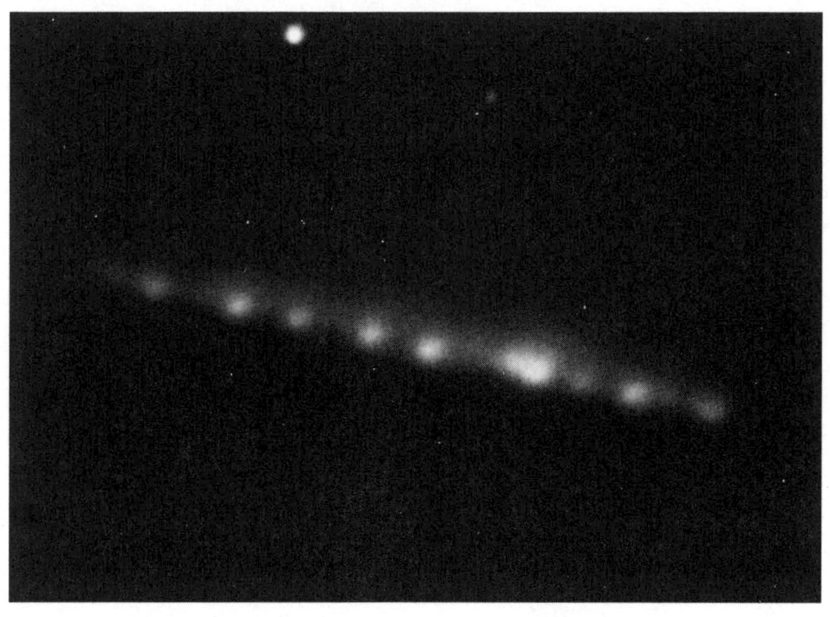

Fragments of Comet Shoemaker–Levy 9 orbiting Jupiter in a row after its breakup due to Jupiter's gravitation and prior to their impact. Courtesy of J. Luu and D. Jewitt.

maker–Levy 9 was ripped apart by the planet's great tidal force into 17 or more fragments, each a mile or two in size, that marched across the sky in close single file, like some ghostly ancient Greek or Roman phalanx. The pieces moved in close formation like the resurrected splinters of the broom that Mickey Mouse, as the Sorcerer's Apprentice in the movie *Fantasia*, was unable to stop. On they went, turning on their captor by taking direct aim at Jupiter itself. On July 16, 1994, the pieces in the line struck one after another, separated by about six to ten hours, with the power of oversized thermonuclear explosions, and releasing as much force as a trillion tons of TNT.

For six days they struck, producing a series of hammer blows, the mightiest convulsions ever witnessed by humans. They produced a ring of dark holes in the upper Jovian atmosphere. The overweight planet rotates once in only ten hours, producing its easily seen equatorial bulge, and thus the wounds appeared widely spaced across its flank. Even the U.S. Congress was impressed by the magnitude of the violence and devastation, and appropriated funds for the study of ways to deflect a possible future onslaught of this kind here on the Earth.

In two successive years, in the spring of 1996 and 1997, we saw two "comets of the century" as dubbed by the media. These were the comets Hyakutake and Hale–Bopp. The first was a rather small affair, but it passed quite close to the Earth. While it appeared bright because it came as close as one-tenth of an astronomical unit (the 93,000,000-mile yardstick of planetary distances), it was small at only a mile or so in diameter, and it flitted by rather quickly. Comet Hale–Bopp, on the other hand, is a colossus among its kind at 25 miles (40 kilometers) in size. It never came closer than 1.3 astronomical units from us, not as close as the Sun, and thus its slow leisurely pace across our sky and its brilliance allowed this comet to be seen by more people

Comet Hale–Bopp. Photo taken by Robert Crelin on April 9, 1997, at South Wallingford, Vermont (15m exposure at f/3.5, 200mm telephoto lens, Kodak Royal 1000). Courtesy of Robert Crelin.

than any other. Had it, too, come as close as 0.1 astronomical units, it might have appeared at magnitude −6, about six times as bright as Venus at its brightest, and would have been very easily visible in daylight. The difference in the closest approaches of Hyakutake and Hale–Bopp had a large effect on the apparent sizes of their tails. Hyakutake's appeared much longer, even if in fact it was not. The visibility of comet tails is, as much as any celestial occurrence, a function of night-sky brightness. From a suburban site in Middletown, Connecticut, I saw as much as 18 degrees of its tail, while at the very same moment, observers at the Van Vleck Observatory on the brightly lit Wesleyan campus, just two miles away, only managed to see the tail 3 to 5 degrees in length. Many observers at truly dark sites reported its tail to be as much as 60 degrees at its maximum.

Hale–Bopp illustrates one other feature of comets in general. They are small in mass, and when they pass by the giant planets, their own orbits can be altered, as was that of Comet Shoemaker–Levy 9. Jupiter, being the most massive planet, does most of the disruption, and it shortened the period of Comet Hale–Bopp during this approach from 4200 years down to 2400 years. Our descendants need wait only about half as long as before to see it again.

Chapter 19

The Seasons

The energy source for the earth and all life upon it is the Sun. The celestial sphere by day is a simple affair. Above our atmosphere, only the Moon and Venus are visible now and then; otherwise the Sun shines alone. Just an average sort of star and gaseous throughout, it is a furnace built of its own fuel. But with 99.8 percent of the mass in the entire solar system, and the next closest star about 270,000 times as far away from us, the Sun is king around here.

Fortunately for us, the radiation emanating from the Sun is very nearly constant and has been so for millions of years. The amount of heat falling on the surface of the Earth depends on the angle of incidence of incoming sunlight and on the length of the period of daylight. It reaches a maximum when the Sun is overhead and becomes insignificant for very large angles near sunrise or sunset. In addition to the diurnal variation of the altitude of the Sun in the sky, a pronounced seasonal variation is present in the incoming solar radiation, at least for high and intermediate geographic latitudes.

The seasons are the result of one major and two minor underlying causes. The dominant major factor is the one that most of us learn in grammar school: the inclination of the Earth's axis with respect to its orbital plane, known as the plane of the ecliptic. This angle, which amounts to about $23\frac{1}{2}$ degrees, in June tilts the Northern Hemisphere toward the Sun and the Southern Hemisphere away from it. In December, the positions are reversed.

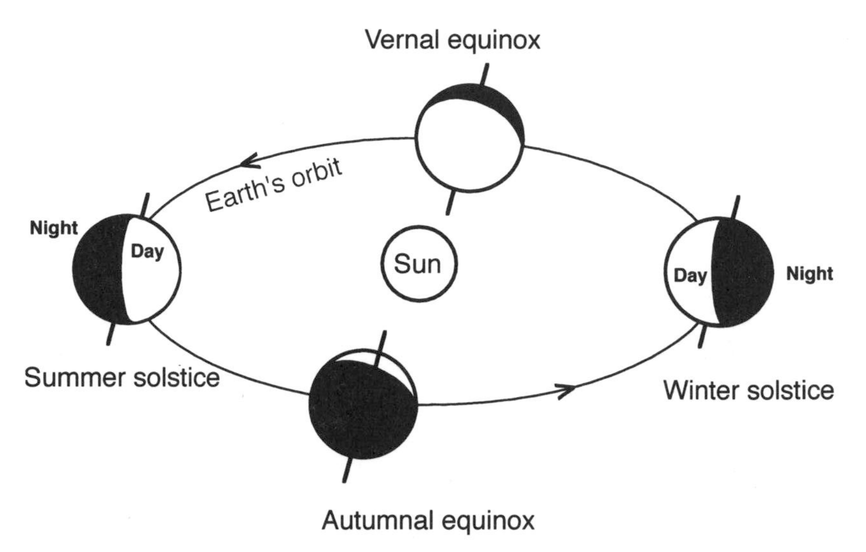

The seasons and their principal cause due to the inclination of the Earth's axis to the ecliptic.

Taken together with the latitude of the observer, the tilt determines the maximum angle of incidence of solar radiation (the altitude of the Sun above the horizon at noon), as well as the duration of daylight. However, extremes in mean monthly temperatures do not coincide with the maximum amount of insolation (a term shortened from incoming solar radiation). In the Northern Hemisphere, maximum insolation occurs around June 21, when the Sun is farthest north in the sky, but the hottest weather of the year normally occurs in July and August. Similarly, minimum insolation is reached on December 21, while the coldest month is either January or February. This effect, the well-known "lag of the seasons," demonstrates the heat inertia of the system of the combined atmosphere and oceans, indicative of the enormous heat capacity of the oceans.

We have already examined the skies of the four seasons. Our society is accustomed to divide the year into four seasons and 12 months. This is no surprise, for our Moon goes through its phases just over twelve times in a year, and the two extremes

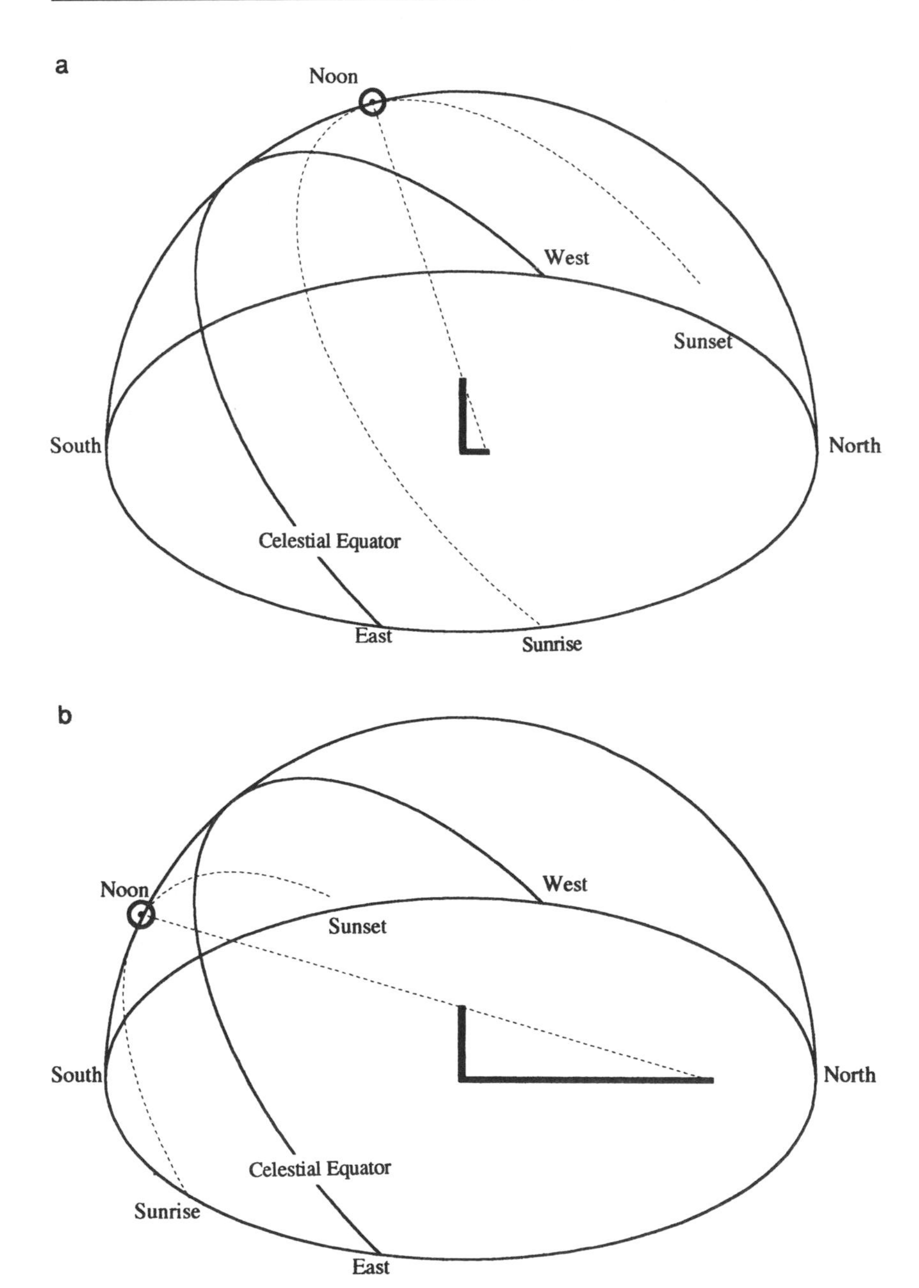

In June (a) the Sun can be seen high in the sky from midnortherly latitudes, but in December (b) it remains low, even at noon.

of winter and summer with two in-between periods make a natural foursome in the temperate climates.

But how are the four seasons to be marked? Which days should constitute each season on the calendar? For the most part, two very different ways of delineating the seasons have been used, with each serving a specific purpose.

The first and best known definition of the seasons is astronomical in origin. It is a natural consequence of inclination of our axis to the plane of the ecliptic. The ecliptic defining the path of the Sun in the sky crosses the celestial equator at two points. The Sun, then, must cross the celestial equator twice each year. It does so on or very close to March 21 and again on September 23, and at these times the periods of daylight and darkness are each 12 hours long as seen from any point on the Earth. These points and dates are known as the equinoxes—the vernal, or spring, and the autumnal, respectively. After March 21 the Sun moves northward away from the celestial equator until June 21, when it reaches the most northerly point on the ecliptic, called the summer solstice. The Sun then moves south for six months until it reaches the winter solstice about December 22. The four key dates, the two equinoxes and the two solstices, determine the astronomical seasons. Holidays marking their occurrence have been part of our heritage for a very long time.

If we look closely at the origin of holidays in the temperate midlatitudes, we can just detect the vestiges of a division of the year into not four, but eight, equal parts. This archaic tradition harks back to prehistoric times, as many ancient peoples shifted from a nomadic hunter-gatherer existence to a settled agrarian one. Agriculture was one of the greatest advances ever made, and it led directly to the rise of villages and later cities and civilizations.

In the temperate regions where most early civilizations arose, early farmers discovered the lag of the seasons. The lag averages about half a season, or a month and a half, more in maritime climates and less in the continental regimes. As a result, the hottest and coldest periods of the year vary in the Northern Hemisphere, but are never far from the first days of August and February, respectively. Halfway between these two dates, on or about the first of May and the first of November, early people marked the midway points at the beginning and the end of the warm half of the year. In many locales these dates correspond roughly to the last killing frost of the spring and the first of the fall, and the period between coincides closely with the frost-free period in many temperate climates.

This is a climatic, or meteorological, manner of dividing the year into four nearly equal seasons. Offset as they are by about half a season from their astronomical counterparts, they give rise to four additional important dates roughly equally spaced throughout the year, subdividing it into eight nearly equal parts.

In late December when the days are shortest, we have the Roman Saturnalia and our own Christmas and Hanukkah, holidays associated with light. Although the time of year of the birth of Christ is not well known, early Christians may have expropriated an old pagan holiday marking this darkest time of the year. With Christmas trees and Menorahs, we celebrate light and hope for its natural return.

Candlemas is widely observed in Europe on February 2. The days surrounding this date are on average the coldest of the year, even though they are longer and brighter than the days around the winter solstice. In ancient Celtic times, this holiday was known as Imbolc. In the United States, February 2 has been converted into Groundhog Day and looks forward to the coming spring. Just after the vernal equinox on about March 21, when day and night are of equal length and spring is on the way, Easter and Passover honor the annual rebirth of life, much as did the pagan holidays before them.

April 30, the night before May Day, has long been associated with Walpurgis Night in Central Europe. As described in Goethe's *Faust* and elsewhere, it is a night of orgy and revelry, appropriate for the start of the half year of warmth, carefree living, and the abundance to come. An unfettered salute to spring, May Day derives from Beltane, an ancient Celtic holiday associated with Druid ceremonies and bonfires.

In Scandinavia and elsewhere, the time near the summer solstice is also a time of unquenchable exuberance. Midsommervaka (midsummer vigil) coincides with the Feast of Saint John the Baptist, held on June 24. The northernmost lands of Europe receive as much as 24 hours of daylight at that time. In the new world, no direct equivalent appears, but many lands, including Canada and the United States, commemorate their independence very soon afterward in early July. Our midsummer gala is replete with fireworks and picnics, far removed from the workaday world and its cares.

And what about the start of August? According to our scheme, this point in the calendar should be marked by a holiday (other than a general period of vacations). None appears on American calendars, but in England, August 1 has been celebrated as Lammas (Lughnasadh in pre-Roman Britain), a kind of early harvest noted for the baking of bread. Lammas night is known as a time for bonfires. The beginning of autumn, falling at the time of the autumnal equinox (about the twenty-third of September, when the day and night are again equal), is devoted to the harvest. Even the Moon becomes a part of the holiday. The full Moon occurring nearest the equinox is known as the harvest Moon. During this time the motion of the Moon is nearly parallel to the horizon, causing it to rise very soon after sunset for several days in a row, providing farmers with extra light for the gathering of crops. The Thanksgiving holiday is unnaturally late, and serves more to introduce the Christmas season than to celebrate the harvest. The English Michaelmas is a much more appropriate holiday for harvest time since it is celebrated on September 29.

Finally we come to the start of the cold half of the year. Unlike the gaiety of May Day and Walpurgis Night that ushers in the warmer half, this time is marked by things that go bump in the night. Known as Samhain by the Druids, All Saints' Day and All Souls' Day (November 1 and 2) are prefaced by Halloween on October 31, our only holiday devoted to the dark side of life, and the only one with black as a predominant color.

* * *

Other astronomical factors have a smaller but still real influence on the seasonal effect. One is related to the distance between the Earth and the Sun. This distance averages about 93 million miles, or 150 million kilometers, but varies during the course of the year due to the eccentricity of the Earth's elliptical orbit. The separation reaches a minimum (of about 91.5 million miles) in early January, and a maximum (of 94.5 million) six months later in early July. At these extreme positions, known as perihelion and aphelion, the Earth receives about three percent more and less solar energy, respectively, than it does on average.

Johannes Kepler discovered that planets move in nearly elliptical orbits and that they move slowest and therefore remain longer near aphelion than they do when nearest the Sun at perihelion. This accounts for the unequal halves of the year. The Sun takes only 179 days to proceed from the autumnal equinox through the winter solstice to the vernal equinox, but it requires 186 days, a whole week longer, for the return trip back to the autumnal equinox. At present the arrival of the Earth at its perihelion point coincides with winter in the Northern Hemisphere and summer in the Southern Hemisphere; when the farthest point is reached, the seasons are reversed. This coincidence has the effect of intensifying the extremes of temperature in the Southern Hemisphere while moderating them in the Northern Hemisphere. But since much of the Southern Hemisphere is covered by water, the intensification there is not very pronounced. This is because the thermal inertia of the oceans tempers the climate; that is, the oceans heat up and cool off more slowly than the land, a feature of maritime climates that causes them to be characteristically less extreme than their continental counterparts. The variation in distance from the Sun and the sizes and distribution of the continents and the oceans are the two minor causes of the seasons.

At the present time the date of perihelion passage in the Northern Hemisphere is in early January, and the aphelion point is reached in early July. But as the equinoxes and solstices are carried backward (westward) along the ecliptic by precessional motion, the passage of the Earth through the closest and farthest points from the Sun will advance in the calendar. In fact, in the thirteenth century, about the year A.D. 1246, these two points

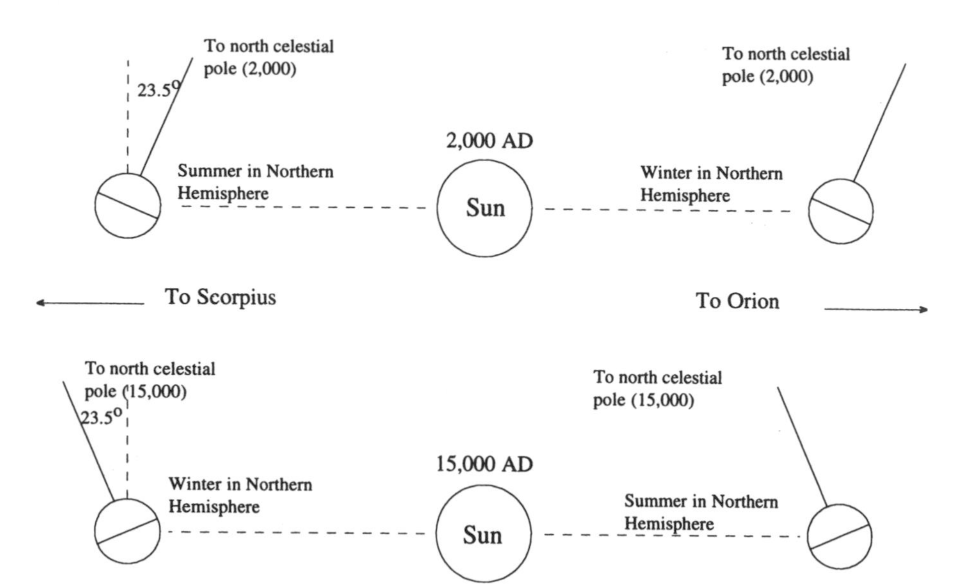

The orientation of the Earth's axis as it is today and as it will be 13,000 years hence.

coincided with the two solstices. Then the perihelion point fell on or about December 22, and the aphelion point on June 22, at the beginning of the summer season.

In 13,000 years all of this will have changed. As a result of precession, the Earth will be tilted—its axis will be pointing—toward the opposite side of the vertical to the ecliptic than it points today, and the extremes of the seasons of that time will be enhanced in the Northern Hemisphere and moderated south of the equator. The great land masses of the Northern Hemisphere do not share the moderate marine climates of their smaller southern counterparts, and the seasonal extremes are expected to be substantially greater than at present.

The angle of the inclination of the earth's axis also varies slightly over the millennia. The present value is very near $23\frac{1}{2}$ degrees, but in 2000 B.C., when Stonehenge was completed, it amounted to a full 24 degrees. The heel stone there, over which the Sun appears to rise on June 21 as viewed from the center of the monument, was set for that earlier and larger value. The angle of the tilt varies from 22 to 26 degrees and back in a period of 41,000 years. This long-term motion, when combined with precession, causes the location of the north pole in the sky and the succession of pole stars to vary slightly. Polaris was near the pole 26,000 years ago, and will mark it again that many years from now, but it was and will be at slightly different distances from the true pole because of the variations in the obliquity of the axis.

A third astronomical influence on the climate of an even longer period arises from variations in the amount of eccentricity of the orbit itself. The present value, causing the annual variation in the distance between Earth and Sun, is now about 1.5 percent, but it will change in a rather uneven manner from nearly zero to some 5 percent over intervals of about 100,000 years. Summer–winter differences may be more extreme when the orbit is more elliptical than more nearly circular, although precession will vary the effect. Cooler summers are thought to be more important in regulating climate than colder winters. Whenever the Northern Hemisphere has the least insolation in the summertime, the accumulated snow of the previous winter may not all melt, and it forms a base on which a new layer of snow can build. Ice ages may begin in this manner.

These three slow variations—precession, the variation in the tilt, or inclination, of the Earth's axis, and the change in the eccentricity of its orbit—taken together have the potential to exert a great influence on climatic conditions and on life. The Serbian astronomer Milankovic was the first to study these long-term effects. He suspected that these slow variations are partly responsible for the cold periods of intense glaciation we call the ice ages, a theory later confirmed by the study of ice core samples.

Other planets have their own seasons. Curiously, all of the planets except Jupiter and Uranus have axial inclinations about the same as ours—about 20 to 30 degrees. Jupiter shows one extreme with a tilt of only 3 degrees, and Uranus exemplifies the other at 81 degrees. Seasons on Jupiter would be nonexistent and those of Uranus extreme were it not for the other astronomical influence—orbital eccentricity. Our figure of only a 1.5 percent eccentricity gives us the third most nearly circular orbit among the planets; only Venus and Neptune have rounder orbits. The orbits of Jupiter, Saturn, and Uranus have near 5 percent eccentricity, those of Mars and Mercury, about 9 and 20 percent, respectively, and Pluto leads with 25 percent. So Jupiter has seasons after all. But unlike the Earth, its two hemispheres would be in phase—both experiencing summer or winter together—when it is near its perihelion and aphelion points.

Chapter 20
About Time

Prior to the mid-nineteenth century, horses provided the fastest, most dependable method for getting around on land. Then came the iron horse, which with the telegraph meant transportation and communication in very short amounts of real time. This marked the first occasion in which the differences in local time brought about by differences in longitude became a major inconvenience.

Each little burg had its own official time, and railway travel, confusing at best, became a hopeless jumble. A train leaving Boston 4 hours and 44 minutes (4:44) behind Greenwich Time (GMT) arrived in New York 4:56 behind GMT, or 12 minutes behind the time in Boston. Travelers had to make the correction to clocks and watches to be on the same time as the rest of New York. Proceeding onward to Philadelphia would require another adjustment of five minutes, because that city is 5:01 west of Greenwich, and Washington, at 5:08 west, meant still another seven-minute change. Railway schedules are tough enough to make sense of without these continual time corrections.

In 1883, the American railroads came up with a way around this mess. They divided our nation into four time zones. All territory between longitudes 4:30 and 5:30 west was combined into a single region with a single time, namely, that of five hours west of Greenwich with an allowance for diversions along state and county lines, as required by state and local ordinances. The time for the entire region was designated Eastern Standard Time.

The next three zones to the west became the Central, Mountain, and Pacific time zones, centered on six, seven, and eight hours west of Greenwich, and later the Canadian maritime provinces were grouped into Atlantic Standard Time.

It is noteworthy that the American railroads tied their time zone concept to the meridian that passes through the Royal Observatory at Greenwich near London. At that time each of the major seafaring nations had its own prime meridian; that for the United States passed through the U.S. Naval Observatory. France observed one that passed through the Observatoire de Paris.

Then at an international meeting held in Washington in 1884, Greenwich became the prime meridian, the standard of longitude for the entire world. Beforehand, alternatives to it were recommended, including the American and French meridians, Rome and Jerusalem for their religious significance, and even the Great Pyramid. But the advantage of a first-class observatory and the predominance of Britain in shipping, where knowledge of longitude is critical, tipped the world toward Greenwich.

Shortly afterward Congress sanctified the arrangement and other governments quickly followed, carving most of the world into 24 time zones, each an integral number of hours ahead or behind Greenwich. A few anomalies can still be found; the island of Newfoundland (but not Labrador) has adopted a time three and one half hours behind Greenwich Time, and India stays at five and one half hours ahead of that time. But most countries of the world are a whole number of hours ahead or behind Greenwich, and almost all of the rest are even half-hours different. This is confirmed at any airport with a display of the local times in such cities as New York, London, and Tokyo. All show the minute hand at identical settings, with only the small hour hands of the several clocks differing from each other.

Most places do not violate their longitudes very much in their adoption of a legal time. But in one extraordinary instance, Stalin was alleged to have considered placing the entire former Soviet Union, then spanning 11 time zones, on the time appropriate for Moscow. Such a brainstorm would prove very unpopular to the folks living in Vladivostok and the rest of eastern Siberia. Their workday started with the Sun well to the west of noon in the afternoon sky, and ended in the wee hours of the night. The measure seems never to have come about for good reason.

The desire for more sunlight in the evening hours after dinner brought about fast time, or daylight time. Whenever applied to Eastern Standard Time, for example, clocks get set one hour ahead, and the time is referred to as Eastern Daylight Time. Then this zone becomes four, not five, hours behind GMT, and differs from Atlantic Standard Time in name only. Similarly, the other time zones take on the time for the adjacent zones to their east.

As the new year sweeps around the world, the difference between the time zones becomes most noticeable. Eastern Asia celebrates its arrival at dinnertime in Europe, and near noon in the Americas. Then it is Europe's turn, followed later by the East Coast and finally the West Coast of North America. Someone has to be first and someone else last. This problem has been solved by the creation of the International Date Line, which completes the great circle that includes the Greenwich meridian. It is thus 12 hours east, and west, of GMT. The line runs through the middle of the Pacific Ocean and crosses little land. Small deviations from 12 hours were imposed to prevent its passage over any land except Antarctica, and until now every nation was almost entirely on one side or the other. But the tiny sprawling island nation of Kiribati, which extends well to both sides of the line, has elected to unite to the west of it, and in the future, the new year will visit Kiribati ahead of all other lands.

Recently the addition (in some years but not all) of an extra second before the arrival of the new year occurs as the time ball descends onto the roof of the former New York Times building and the crowd in Times Square goes wild. As the ball drops, the second after 11:59:59 P.M. is not followed by 00:00:00 (midnight) but by 11:59:60, another second of the departing year.

What is the reason for this? It results from the fact that the Earth is a poor timekeeper. The Moon and the Sun tug on it as they will, and in doing so, form the tides. The tides slosh the oceans around so as to brake the Earth's rotation. Over the billions of years since their formation, they have slowed the day from less than 8 hours to the present 24. As time passes the day will lengthen further until we keep the same face to the Moon as it now does to us. Scientists, in their most precise work, have replaced the time kept by the erratic planet Earth with resonations of the Cesium atom, which, by international definition, makes 9,192,631,770 resonations a second. This precision, adapted to the longitude of Greenwich, is called Universal Time (U.T.) to distinguish it from the less evenly paced GMT, which may be a fraction of a second off. That extra second added between the years accounts for the effects of the tidal slowdown, and places the year right back where it belongs. Sometimes the extra second falls at the end of June, since irregular fluctuations in the rate of rotation occur as well as the tidal deceleration.

Other properties of the solar system lead to other difficulties with time and the calendar, with adjustments needed to keep things (like the arrival of spring) on their proper schedule. Julius Caesar was one who dealt with the irregularities of the calendar; in order to stem local corruption and bring uniformity to the Roman Empire he asked a Greek astronomer named Sosigenes to determine a calendar of some reality and regularity. The year was known to be about $365\frac{1}{4}$ days long, and this exact length was adopted in 46 B.C. Since four years amount to 1461 whole days, 1 whole day was added to February every fourth year.

The year, however, is not *exactly* 365.25 days long, but about 11 minutes, or 0.0078 days, shorter, making it 365.2422 days in length. Every four centuries, things would get out of whack by just about three days. So by the time the Middle Ages arrived, each season started earlier. By A.D. 730, Bede, the learned English monk and scholar, knew that the Julian year was too long by 11 minutes and 14 seconds. It took another eight centuries for anyone to do anything about the problem. In 1582, when Pope Gregory XIII acted to correct it, the error had accumulated to ten days, not from the time of Caesar, but from the time of the Council of Nicea about A.D. 325, when the newly Christianized Roman Empire adopted the Julian calendar.

Pope Gregory summarily dropped ten days from the calendar; thus, Thursday, October 4, 1582, was followed by Friday, October 15, of the same year. This change brought the date of the

October 1582

S	M	T	W	T	F	S
	1	2	3	4	15	16
17	18	19	20	21	22	23
24	25	26	27	28	29	30
31						

The calendar month of October 1582 shows Thursday, October 4 followed by Friday, October 15, as the Gregorian calendar took effect by dropping ten days out of the month.

vernal equinox from about March 11 back to its customary date of March 21, and similarly for the other primary dates.

One more correction was needed to prevent further hemorrhaging of time in the future. Each future 400-year period needed to be shortened by three days. The days chosen were the leap year days falling in the century years not divisible by 400. Thus 1600 was a leap year, but 1700, 1800, and 1900, which were leap years in the Julian calendar, were not leap years under the new system. The year 2000 is a leap year in both, but 2100, 2200, and 2300 will not be in the Gregorian calendar.

The Roman Catholic countries adopted the new Gregorian scheme at once, but the others wanted no part of the papist business at first. By 1700 most European countries had gone along, and Britain finally did so throughout its empire in 1752. By that time 11 days were needed for the transition, and so September 2 of that year was followed by September 14. George Washington was born on February 11 in the year 1732. But the date of his birth shifted to February 22, 1732 as a direct result of the transition. Non-European nations slowly followed suit, and this century marks the first in which the entire world reckons by the same calendar. With the adoption of the Gregorian calendar, the British Empire and others also adopted the practice of beginning the year on January 1, instead of March 21, as had been customary beforehand.

I stated earlier that the year accounts for 365.2422 days, but three days dropped out over four centuries is equivalent to a year of 365.2425 days. A discrepancy of 0.0003 days per year, or around one day every 3000 years, is still there to confound our descendants. Sometime before A.D. 3000 or so, the United Nations or what passes for it at the time must act again to fine-tune the calendar to adjust for even this small shortfall of time.

The calendars discussed here are known as solar calendars because they are founded on the rotation of the Earth along with its motion about the Sun. The Moon does not enter into the picture in either of the two principal calendars of our Western culture. But in early times, there were lunar calendars, making use of the length of a lunation or synodic month, whose average duration is 29.531 days. Twelve of these periods amount to 354.372 days, short of a year by 11 days.

Means to reconcile the lunar and solar years were sought; the most successful make use of the fact that 235 lunations last almost exactly as long as 19 Julian years. This period is called the Metonic cycle, after Meton, who noted its earlier discovery about 433 B.C. The relationship means that the phases of the Moon recur, after 19 years, on the same days of the month, with an occasional shift of one day. The Metonic cycle plays a part in the ecclesiastical calendar for finding the appropriate date for Easter. Easter and the other movable feasts that depend upon it can shift over a period of more than a month. In the Western church, it is fixed as the first Sunday after the first full Moon following the Vernal equinox. Differences are still found among other churches, but these no longer lead to warfare, as they were wont to do in the early Middle Ages.

A number of calendars of Near Eastern lands, including the Hebrew calendar, use this arrangement by intercalating 7 extra months to the 228 months that occur over an interval just about equal to 19 Julian years. Thus 12 years have 12 months and the other 7, 13 months. These are known as lunisolar calendars. In the lunar calendars of many Islamic cultures, the month began with the first sighting of the new Moon. They are mostly desert lands with a high percentage of clear weather, where the thin crescent Moon can frequently be spotted less than two days after the moment of new Moon. Under the best conditions the very thin crescent is seen even less than one full day after the new Moon by trained observers.

CHAPTER 21

Ellipse

Johannes Kepler published his first two laws of planetary motion in 1609 and his third law ten years later. With them he revolutionized astronomy and gave the solar system its modern shape. All before him, including his great contemporary Galileo, sought to portray all motion as circular in form. Even Copernicus and Galileo, who held that the Earth does actually move in rotation about its axis and in revolution about the Sun, championed the circularity of orbits as a form of the most perfect shape.

Kepler tried again and again to make the orbits circular, using the most extensive and precise observations of his former master, the Danish astronomer Tycho Brahe. But no combination of circles could be made to fit the data. Ushering in a tenet of modern science, he required his model to conflict with the data within the uncertainties he perceived. With ellipses he found the reconciliation he was looking for. His first law maintains that all planetary orbits are elliptical in form, with the Sun occupying one of the two foci (see the figure on page 198).

An ellipse is a closed plane curve generated by a point moving so that the sums of its distances from two fixed internal points, known as foci, are equal. Each focus is always found along a line marking the largest diameter, or major axis, of the figure, which lies perpendicular to its smallest internal diameter, or minor axis, and intersects it at the center, midway between the foci, as shown in the figure. Either half of the major and minor

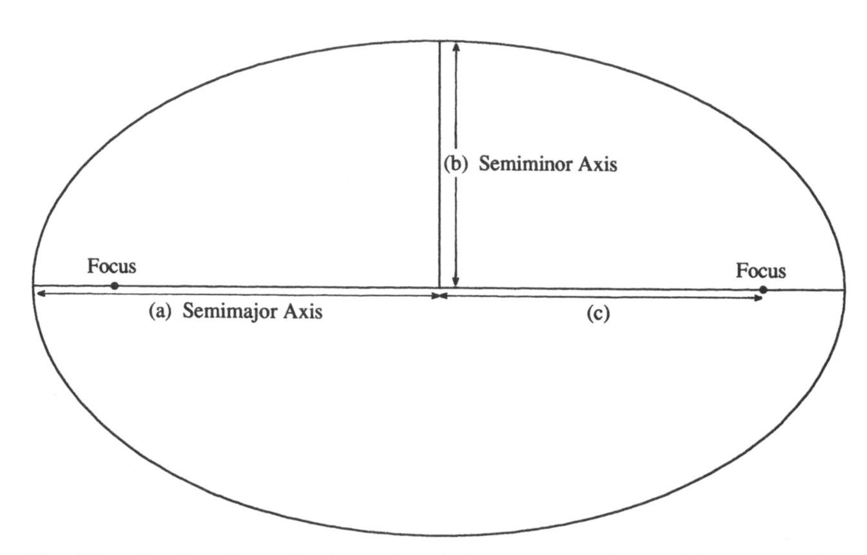

The ellipse showing the semimajor and semiminor axes, a *and* b, *and the distance from focus to center,* c. *The sum of the distances to the two foci is the same for every point on the ellipse. The eccentricity is defined by the ratio of* c *to* a.

axes are designated by a and b, respectively. The quantity a, then, is known as the semimajor axis, or the semiaxis major, or even the major semiaxis, names raised among the plaints of a vocal minority, in a syntactical dispute that remains unsettled. The semiminor axis, or b, fares no better in this business. One more parameter needs to be defined; this is the distance from either focus to the center, known as c. The ratio of c to a defines the eccentricity, e. Thus $e = c/a$. Together with a, e fixes the size and shape of a planetary orbit. Ellipses with constant size, a, and different eccentricities, e, are shown to the left in the figure on page 199, while the reverse appears to the right in the same figure. At e = zero, the two foci converge onto the center and you have a circle. At the other extreme, whenever e is equal to or greater than unity, the result is an open-ended parabola or hyperbola. Since the eccentricities of the planetary orbits are small, miscasting them as circles was easy, and they are near-circles in point of fact.

Kepler went on to show in his three laws of planetary motion that future positions of the planets could be predicted far better

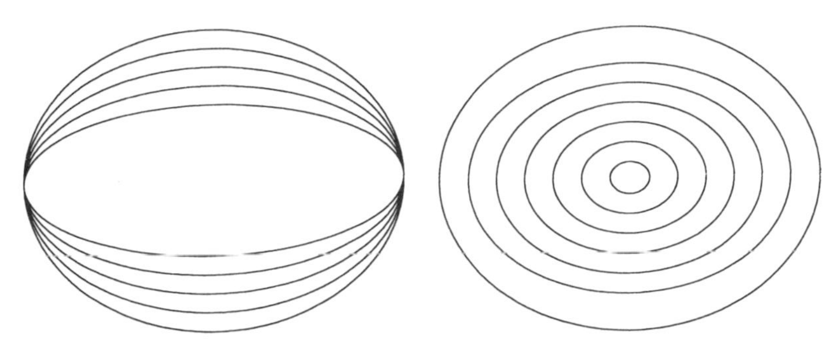

Ellipses with constant a *and variable* e *(left) and constant* e *and variable* a *(right).*

than they could in any system of circular orbits. This concept, the ability to predict, has been realized as one of the primary tests for any new hypothesis in competition with older accepted ones.

Kepler was the first to discover the harmonic law, his third law of planetary motion. It related the period of each planet, the time it takes to orbit the Sun, to its mean distance from the Sun and by so doing tied the entire solar system into one entity. He found that for each planet, the square of its period in years is equal to the cube of its average distance from the Sun, measured in terms of the Earth's average distance, defined as one astronomical unit. It remained for Newton to generalize this law by incorporating the masses of the two orbiting bodies into the equation. By doing so, he extended it to apply throughout the universe.

The great power in this law lies in its ability to allow masses to be determined. If there is one single intrinsic property of a celestial object that an astrophysicist wishes to measure, it is its mass. Mass is the property that determines whether a condensing nebular cloud will become a star, an earthlike or terrestrial planet, or a major or Jovian planet. It's all in the initial mass. Newton's generalization means that if a satellite, whose own mass is relatively very small, orbits a planet (or a larger satellite such as the Moon), we need only find its period and its average distance away from the body it is orbiting, and, bingo, we have its all-important mass! Artificial satellites have provided the masses of many objects of the solar system in just this way.

Not only has this provided the touchstone upon which the physical nature of the members of the solar system rests, but it is the one way by which the masses of other stars can be derived directly. The chemical composition varies only a little from one star to another, but the mass can and does vary widely. The mass of a star determines its luminosity, its surface temperature and therefore its color, and its lifetime, or how long it will be a star at all.

The two stars whose light merges to form the Capella or the Alpha Centauri our eyes see, by revolving around each other, give us their masses. We can only compare a single star to a similar component of a pair to estimate its mass. The study of double and multiple stars is a very important field, because they alone can yield the masses of stars.

Chapter 22

A Tale of Two Stars

It is therefore a truism, almost a tautology, to say that all magic is necessarily false and barren; for were it ever to become true and fruitful, it would no longer be magic but science.

—J. G. Frazer
The Golden Bough

Arcturus and Vega make a nearly ideal pair of stars to illustrate the differences in the physical properties of the stars in general. As we have seen, these two brightest stars of summer appear at a magnitude of just zero, and both are at about the same distance from us (Vega is actually a little closer). That means that intrinsically they are also almost equally bright.

Observers can use the two stars to test the color response of their own eyesight. On any clear, moonless night in the spring or summer, look at the two stars when they are equally high in the sky. Make sure no clouds can be seen. Except in the remote country, lights will illuminate even the highest clouds and render them visible. Which star appears brighter? Some contend that the blue Vega outshines Arcturus, whereas others see the orange star as the brighter. Photoelectric photometers, devices that measure light with very high precision, find that Arcturus sends us just a bit more visible light, but this difference is much too small to be detected by the eye or by the camera. Those who see Vega as the brighter star probably have eyesight that is slightly more sensitive to blue light. Others who see a little more red light may find Arcturus the brighter star.

A glance shows the big difference between them to be their color, which is dependent on their surface temperatures. The coolest stars appear reddish in color, and as hotter stars are considered, the colors appear orange, yellow, white, and finally blue for the hottest of all. The Sun appears yellow and has a middling (for a star) surface temperature near 6000 degrees Kelvin (degrees of the Celsius scale measured from absolute zero, such that 0 degrees Kelvin equals −273 degrees Celsius). Arcturus is a little cooler and Vega quite a bit hotter, or about 5000 degrees and 10,000 degrees Kelvin, respectively, in surface temperature (the interior temperatures of all stars are much hotter).

A well-known relation of physics connects a star's luminosity and surface temperature with its radius, or size. Luminosity varies with the square of the star's radius and the fourth power of its temperature. That means that Vega, being twice as hot, must put out about 16 times as much radiation per unit area as does its cooler neighbor. Therefore Arcturus must have 16 times as many square miles of surface area or it would not be as bright, and we know that the two stars are nearly equal in innate brightness. It must be about four times the diameter of Vega, a giant in every sense of the word. The diameter of Vega is almost three times that of the Sun, making the orange giant over a dozen times the Sun's size. Arcturus would fill a fair portion of Mercury's orbit were it to displace the Sun.

These two bright lights don't even come close to the limits of stars in size, brightness, or temperature. The history of stellar astronomy in this century is dominated by the ever-increasing ways in which the true physical properties of stars can be predicted by their easily measured features. From apparent brightness and distance we derive the true or intrinsic luminosity. From the relation between luminosity, temperature, and size, we obtain each quantity in terms of the others and in absolute physical units. And from a number of considerations we are able to determine directly or indirectly the mass of a star. With this information the whole history of a star can be deduced, as can the process going on in its interior.

From surveys encompassing millions of stars, astronomers have a clear picture of the sizes, colors, and luminosities in which stars can be found. Compelling evidence reveals that a close association is found between luminosity and color, or surface temperature, for most, but not all, stars. A graph of this relation appears in the figure on this page. Graphs of this kind are called Hertzsprung–Russell (abbreviated HR) diagrams after the two astronomers who first made them shortly before the First World War. Plots of many stars on HR diagrams reveal the close correla-

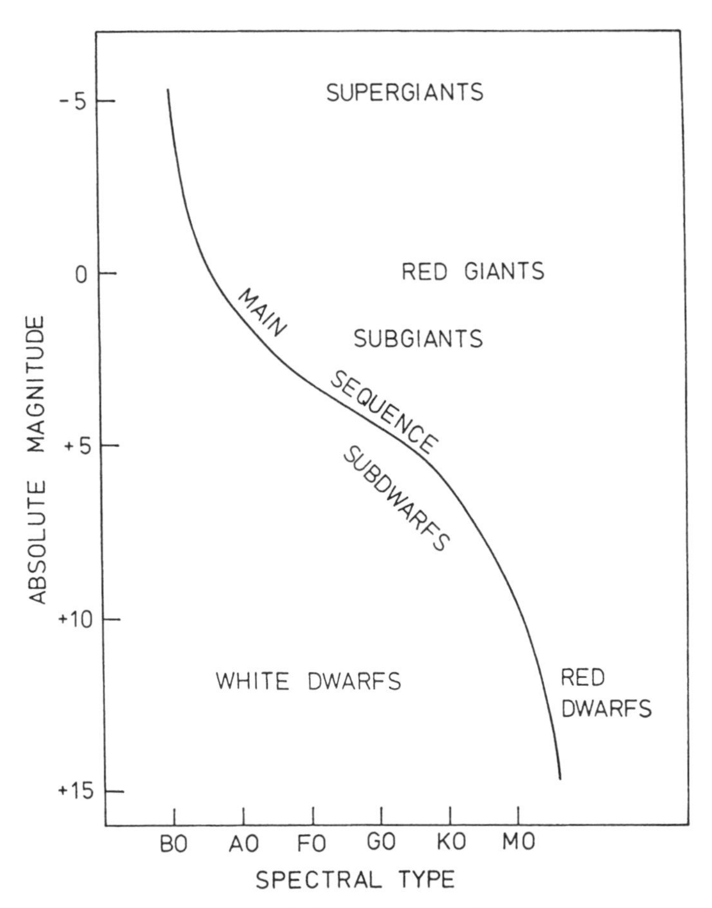

The HR diagram shown schematically. The regions inhabited by the main sequence stars (normal dwarfs), giant stars, and supergiants are shown.

tion known as the main sequence, which we now know represents the position of a star throughout most of its life. But a minority of stars lie above and to the right or below and to the left of this main sequence.

The first of these groups resembles and in fact includes Arcturus. These stars are all relatively red and cool, yet they are many times as bright as the Sun. Like Arcturus they must be giants in order to shine as they do. The second group is just the opposite. These stars shine at hot temperatures and appear blue or white in color, yet are intrinsically very faint. Hence they must be very small. We know that they range in size about as the planets do, with many as small as the Earth. These stars are called, appropriately enough, white dwarfs.

A few stars range along the top of the HR diagram above the giants. These are the supergiants. Most lie to the right of the top, or bright end, of the main sequence and all are among the brightest objects in our galaxy. Some, like blue Deneb and Rigel, are not far from the main sequence, and some, such as Betelgeuse and Antares, are found in the upper right corner of the diagram. They are red, cool, and extremely bright. We know from this and other evidence that they are indeed about the largest objects of all. Were Betelgeuse placed where the Sun is located, the orbits of Mercury, Venus, the Earth, and Mars would all lie within its surface; in fact it would fill the entire inner solar system all by itself! Although hundreds of light-years away, it appears to subtend the largest apparent angular diameter of any star except the Sun. It is also about the largest object we can see easily in the night sky; the Moon is the smallest. The ratio of their diameters is about 200,000 to 1. This ratio equals that of the Moon compared to an object about 50 feet in diameter, the size of a large house or barn.

At the lower right of the HR diagram are the faintest stars of all; they are cool and red, but shine so feebly that they appear not much bigger than Jupiter. These are the red dwarfs, the most abundant stellar life in the universe. Despite their numbers, not one can be seen with the naked eye. Here we come to the most striking fact of all, the central feature of this and all other galaxies. The relative abundance of stars, the number found in a particular volume of space, is nearly inversely proportional to luminosity. Space is filled with little red dwarfs but the giants are rare, and the supergiants much more so.

This phenomenon has an analog in human fame in the ratio of less well-known citizens to celebrities. A survey of the names that are most frequently found on the front pages of newspapers would turn up mostly major public figures: politicians, entertainment and sports figures, and the like. In contrast, a survey taken of the people living in your own neighborhood is very unlikely to turn up even one of these celebrities. But the names of common folk of no individual fame would predominate, probably unanimously.

We look out into space and see the denizens of the upper main sequence, the giants and the supergiants, in abundance. But for each one of them thousands of red dwarfs and even dozens of white dwarfs are shining in between. Our Sun is an average yellow star sitting right in the middle of the HR diagram. About 90 percent of all stars are fainter than the Sun, yet of the 500 stars brighter than the fourth magnitude in the entire heavens, only two belong to this intrinsically fainter group. One, a star of magnitude 3.5, is located in the constellation Cetus. Not far away lies the other, a slightly fainter star in Eridanus, the long river winding from Orion to the extreme southern sky.

What drives the stars? Astronomers and astrophysicists have devoted much research to the physical processes that produce the stars' untold amounts of energy over millions and billions of years. In the last few decades we have developed a very clear picture of the life history of any common star. The process that brought this knowledge has been likened to the scrutiny of a single snapshot of an extended human family or of a forest, with the intent of inferring from it how babies grow to become children and then adults or how an acorn becomes a seedling and in time a giant oak tree. Unlike people and trees, however, stars have not changed their appearance or physical characteristics in all of recorded history, with very few exceptions. We see them as we have always seen them, at only one epoch in their long lives.

We know that all stars are made primarily of hydrogen, the lightest and simplest of all of the chemical elements. The remaining material consists mostly of helium, the second-lightest and second-simplest element. All of the remaining elements make up at the most only three percent of the material forming the stars. Since all stars have nearly the same chemical makeup, some other factor must account for the tremendous differences that are found from one star to another.

As stars condense out of interstellar gas and dust, they become hotter; their gravitation pulls them together and their cores heat up to tens of millions of degrees. At such temperatures, their plentiful hydrogen is converted into helium by the process of thermonuclear fusion, the same process that fuels a hydrogen bomb. Think of a star as a great gaseous furnace composed of its own fuel in the form of hot hydrogen gas. In the fusion process, each four hydrogen atoms form one helium atom but there is a little mass left over. Einstein's famous equation, $E = mc^2$, indicates that a little mass can become a lot of energy under certain conditions. This is the only process that can keep stars burning for millions of years.

Astronomers are well aware now that the history of a star is determined predominantly by its mass. When a star begins to condense out of a cloud of interstellar nebulosity, the mass it contains determines its future life history. A star like the Sun will quickly (for a star) condense onto the Sun's place on the main sequence, whereas a star of two or three times the solar mass will locate itself somewhat farther up the main sequence; that is, it will become a brighter star. Sirius and Vega are examples of stars of this greater mass, and they are 30 to 50 times as luminous as the Sun. Luminosity is a good measure of the total energy given off by a star, so it appears that they are likely to run out of their mass much more quickly. The enormous energy given off in this fusion process continues in the Sun's case for over 10 billion years. Since it is only 4.6 billion years old, it has more than 5 billion years to live. Stars like Vega and Sirius will last only about a tenth as long; they may live only half a billion years. And Deneb or Rigel may not last for 10 million years at the rate they are burning their hydrogen.

What of the little guys, the small red dwarf stars that dominate any stellar census? They too use fusion to shine but they do so at a slow rate, so slowly in fact that all of them that ever formed are still merrily shining now as when they were first born out of the primordial ooze. Our galaxy is somewhere near 15 billion years old, some three times the age of the solar system. Even at that age, the little red dwarf stars show no signs of evolution from their primary state.

But surely this range in stars must have an end. The planets are not now nor have they ever been stars. They never shone by their own light at any time and they never will. Jupiter, our largest planet, is also mostly made of hydrogen and helium just like the Sun. But Jupiter has only about one thousandth the Sun's mass. It has the right constituency to shine by fusion, but it is too small to build up the enormous central temperature necessary to start the fusion process going in the first place. Somewhere between the Sun and Jupiter must be a borderline mass marking the smallest amount of matter able to trigger the fusion of hydrogen and become a true star. The borderline seems to be near eight percent of the Sun's mass, or 80 Jupiters. Just below this critical mass limit an almost-star may actually shine for a while just from the gravitational collapse from a nebula into a star. But then it will cool off and become a kind of large planet.

Astronomers measure distances within the solar system not in miles or kilometers, but in astronomical units, in units of the average distance between the Earth and the Sun, that 93,000,000-mile distance many first learned in grade school. Interstellar distances are often given in light-years (see Chapter 4). By an interesting coincidence, the ratio between the a.u. and the light-year is just about the same as that between an inch and a mile, 1 to 63,360. If the distance to the Sun were scaled at one inch, light-years would be measured by miles.

At this scale, the Sun would be about one hundredth of an inch in diameter, about the size of a period or the dot over the letter *i* or *j* in small print. The planets would be dust motes or smaller and would require a magnifying glass to see. Jupiter lies some five inches from the Sun, and Pluto about forty. The whole planetary system could fit on a round dining room table seven feet across.

How far away are the stars in our model? The nearest, the triple star Alpha Centauri, would be two dots nearly 20 inches apart and 4 miles away. The third, smaller star orbits the pair at a distance of 300 yards. Sirius is a mustard seed at 9 miles and Vega is another, 25 miles off. At 37 miles a pea represents Arcturus, and Betelgeuse is a grapefruit between 500 and 1000 miles away. Within 30 miles of the Sun in all directions, we would find only about three or four mustard seeds and several dozen periods, a few of which may be surrounded within a few feet by dust motes. That's all. Empty space is very empty indeed.

* * *

More than half of all stars are double or multiple, physically connected and gravitationally bound in orbits about each other. The Sun, as a single star, is in the minority in this respect. Double stars, two stars circling a common center of gravity or barycenter, are found in every possible combination within the assortment of stars in general. Triple stars are less common, but not rare. A few quadruple and quintuple stars are known to exist, as are two sextuple stars. In almost every case, the members of multiple stars or star systems are paired with wide distances between one pair and another, or between a pair and a third single star orbiting the two.

Thus the triple system Alpha Centauri, the closest to our solar system and the third brightest star system in the sky, contains a star that is a near-twin of the Sun and a second star a little more orange, cooler, and fainter, separated by about 20 astronomical units. The distance between them is similar to the separation of Uranus from the Sun. A third star, a very small, faint red dwarf star, lies some 10,000 astronomical units from the pair, or one-sixth of a light-year. The components of a stellar system are lettered in order of decreasing luminosity; thus, the sunlike brightest star is Alpha Centauri A, its closer companion is B, and the faint, red, distant companion is C. Another name for C is Proxima, because it is a little closer to us than the pair AB; it is the closest known object of any kind to our solar system but yet not part of it. Closer than any star but the Sun, it is so faint that it can be seen only with a telescope of about four or more inches in aperture. It is one of very many faint red dwarf stars, shining like ten-watt light bulbs in our Milky Way galaxy. The two stars A and B orbit each other with a period of 80 years, and Proxima trips around the two of them in perhaps a million years.

Suppose the Sun had an orange companion at the distance of Uranus, as well as a faint starlike Proxima one-sixth of a light-year away. What would our solar system look like? First of all, the planets from Jupiter out would not be likely to exist at all. Their orbits would be unstable due to the large gravitational perturbation of the stellar companion. Second, and worse for us, B would shine at the Earth with an intensity of about a thousandth that of the Sun. This doesn't sound like much, but we must recall the eye's ability to account for substantial changes in light intensity. For example, on a clear day, the sky overhead at sunset is only some one-thirtieth as bright as it was at noon. And at civil twilight about a half hour later (about the time when headlights are necessary when driving), the brightness of the zenith is diminished by another factor of 3000! Few would guess that the drop is this excessive. So the solar companion B would by itself create a bright-twilight or near-daytime sky whenever it was above the horizon. When B was seen in the direction of the Sun, the sky would resemble the one we see, but six months later B would appear opposite the Sun in the heavens, and we would have continual daylight all over the world. When both suns were below the horizon, we would have night, and if the Moon was in the sky it might appear at the quarter or gibbous phase, but with a ghostly second crescent on gibbous light, orange tinged, overlapping the first.

Meanwhile, little Proxima would appear all but stationary, and in the night sky would be seen as only of the fourth magnitude, just visible to the unaided eye on a decent night.

What if we had one Sun, but two moons? We think of our satellite with a monotheistic tendency as we do the Sun, but with a second, even our language would change. Full Moon? Which full moon? There might be two of them, or the other might be in the crescent phase. Certainly the closer would eclipse the other at times, and eclipses of the Sun would be more frequent events. Songwriters would have a time of it, and any werewolves might become schizophrenic. Only by chance would the two appear of the same angular size, and if we had three or even four large ones as Jupiter does, our sky would be full of phases, eclipses, moonrises, and moonsets, and the tides would be in a frightful state of complexity. The *Old Farmer's Almanac* would need an addendum just to keep up with it all, and favored times for fishing might as well be discarded altogether.

When Nicholas Copernicus put forth his heliocentric system, he and others realized that the stars should reflect the motion of the Earth around the Sun as we now know they do. From Aristotle until well after the time of Copernicus, many tried to observe this stellar motion called parallax. All failed. Using our astronomical model, they would have had available a baseline of two inches, the diameter of Earth's orbit, to triangulate on little dots miles away. We all have a convenient example of a two-inch baseline since this is just a little smaller than the average inter-

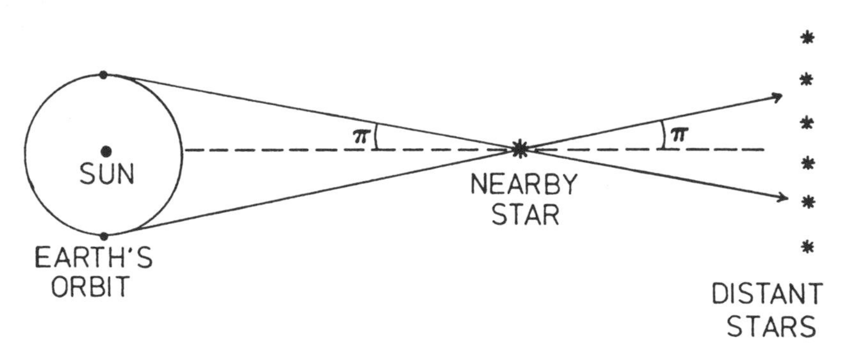

As the Earth moves around the Sun, nearby stars can be seen to move back and forth against a background of distant stars.

pupillary distance, the separation between our eyes. Open one eye and then the other to see objects move; the nearer the object, the farther it appears to shift its position. But who could see and measure the angular shift of a period or a pea miles away?

And yet the stars must move if the Earth is to move. Long after Galileo, Kepler, and Newton had given us a new universe, observers still sought to measure the distances to the stars by their parallaxes. One attempt after another brought about a number of serendipitous discoveries but no parallax. Finally in 1838, three centuries after the Copernican model was published and over two centuries after its wide acceptance, instrumentation caught up with theory. Three astronomers, each working independently of the others, each determined an accurate distance to a star from its parallactic motion.

What happened in the meantime to convince scientists that the Earth moved? In the century after Copernicus, more and more scientists accepted him over Ptolemy as they came to realize that stars were not faint little planet-sized things just beyond Saturn but suns equal to ours and therefore very very much farther away than any planet. The exact course of this paradigm shift is not fully documented even now, but it is surely one of the greatest in all of human thought. It divorced forever the realm of the stars from the solar system, and they could never again be taken for mere adjuncts of either the Sun or the Earth.

This new view of the stars also introduced a change in the way we do science. In the Middle Ages, a concept required proof in every particular to be accepted, especially if any part of it contradicted the prevalent doctrine of the day. After Galileo, a theory need only cope with the observed data better than its competitors to be accepted as a working model. The Sun-centered system never vanquished its competitor in a single blinding stroke of enlightenment. Instead, evidence—direct and indirect—steadily whittled away at tenets common to the pre-Copernican systems, and they faded away in the seventeenth century.

Since then, scientists have accepted the work of Newton, Einstein, Darwin, and Freud among others, not because they answered every lingering question, but as hypotheses serving as models or frameworks on which to build, even though they were not unequivocally proven. None may be the final word even now, but all have proven far superior to the science they replaced. Albert Einstein summed it up very well when he said that "the important thing is not to stop questioning."

Chapter 23

What Ayla Saw

Within the last turn of the great year, the sum of written history has passed and much much more. Twenty-six thousand years ago, Neanderthal man was just dying out (or may have interbred with newcomers, although a new study refutes this hypothesis), and our ancestors, the Cro-Magnon people, had moved into Europe from Africa not long before. The most recent period of glaciation, called the Wisconsin Ice Age in America and the Würm Ice Age in Europe, was in place, with an ice cap spread across the northern half of Europe, Asia, and North America. A small portion of southwestern Wisconsin remained a kind of island mostly free of glaciers, hence the name.

This was the world of Ayla, the heroine of Jean Auel's bestseller *Clan of the Cave Bear* and the other books of her Earth's Children series. Ayla's world occurred during an interstadial period, a time when the ice age let up a little, before its coldest period about 10,000 years later on. It was not warm enough to be classed as an interglacial period, similar to that of the last 10,000 years, when forests of oak and other deciduous trees moved in to cover Europe. Rather, this was a kind of temporary easing of the intensity of the cold and the ice fields of that age.

What would Ayla have seen as she pondered the night sky, assuming she lived close enough to 24,000 B.C.? Polaris was then the pole star, just as it is in our time, one full oscillation later. It would have been a little farther from the pole at its closest than it is now, because at that time the tilt of the Earth's axis was just over 22 degrees, more than a degree less than its present value.

The major visible difference, perhaps the only one, between her sky and ours, would be distortion in some of the recognizable constellations. The Big Dipper would show a putative distortion from its shape today, but would still be recognizable, as shown in the figure on this page. Orion and Scorpius would be only slightly warped since they are composed of more distant stars that move much more slowly across the sky. A handful of fast-moving bright stars would be badly misplaced and it wouldn't require an astronomer to spot them; Arcturus would be situated in northern Boötes about 16 degrees north of its present location, and nearby Sirius and Procyon would each lie some 9 degrees north of the points where we see them today. One precessional lap into the future will displace them by this much again, as Polaris next takes its turn at the pivot of the sky in A.D. 28,000.

This distortion of shape is caused by proper motions, the individual motions of the stars across the sky. We know the Sun to be moving at a speed of 12 miles, or 20 kilometers, per second (amounting to four astronomical units per year) toward Vega, the brightest star in the summer sky, and away from Sirius, the brightest of all, located nearly opposite Vega on the celestial sphere. The rest of the solar system is carried along with it, sharing this motion on top of all the others. The Earth orbits the Sun with a heliacal motion, spinning about it at 20 miles, or 30 kilometers, per second, while sharing the Sun's linear motion of two-thirds that speed.

The Big Dipper today (left) and as it appeared 26,000 years ago (right). The two stars at the ends move differently from the five stars in between, which move as a group.

Every other star has its own individual motion, moving along with respect to its stellar neighbors. Each moves randomly in speed and direction, each acting like one of a swarm of gnats, moving in its own fashion with respect to the other gnats. The apparent angular motion through the sky is the proper motion, measured in seconds of arc per year. Due to the great distances separating them, they cannot be seen with the naked eye to change their positions in a human lifetime. But in a longer interval the small proper motions will cause an apparent displacement of the closest stars with respect to their background constellations, and eventually they will erase the familiar star figures altogether. In time all of the constellations and asterisms will lose their identity and their shape.

The proper motion was first noticed by Edmund Halley about the year 1700, as he was comparing his star positions with those of Hipparchus, made in Alexandria almost 20 centuries earlier. Halley discovered that a few bright stars had shifted their positions within their constellations, even after the effects of precession were removed from consideration. Sirius, Procyon, and a few other bright stars had moved by almost one degree, and Arcturus had covered more than a degree during the interim. He concluded correctly that stars are loose in space and not bound to the surface of a celestial sphere. If confirmation had still been needed that stars were distinct suns at different distances, this discovery proved it beyond a doubt.

Edmund Halley could not have observed the bright star that moves faster than any other. This is the closest star (or star system) to us, the triple star Alpha Centauri, only four light-years away. We have noted that it is the third brightest in the heavens, and appears yellow, much like the Sun. Although not a fast moving star system, its proximity renders its apparent proper motion to be rapid, much as a nearby firefly may appear to move faster than a distant airplane. It covered a very conspicuous two degrees between the times of Hipparchus and Halley. Although Halley's predecessor could just see it, the star lay much too far to the south to be seen by Halley, invisible even in southern Europe after 3000 B.C., when precession carried it below the southern horizon about the time that written history began. In the last great year, it has traveled more than 26 degrees westward across the deep southern sky.

Of more interest than the past is Alpha Centauri's motion in the future. It now lies four degrees east of blue Beta Centauri, the eleventh brightest of all stars, and the two make a more spectacular pair even than Castor and Pollux, the twins of Gemini. Alpha is taking a bead on the more distant Beta, and 4300 years from now it will approach within one-quarter degree to the north of it (less than half the apparent diameter of the Moon) forming a spectacular naked-eye double star. But alas for us poor Northerners; by the time precession carries the two stars back into our ken, Alpha will be well to the west of Beta. Then it will move on, and 12,500 years after that, it will stand at the head of the Southern Cross. This event will be visible to our descendants in northern Europe and North America, as the cross is brought northward again.

Over the length of our written historical record only small distortions have taken place, because only a few nearby stars would appear displaced over the interval. The pyramid builders of almost 5000 years ago would still feel at home under today's night skies. Their sky would rotate about the star Thuban, but the skies in Ayla's distant time, 260 centuries past, would spin around Polaris as they do now.

It will take hundreds of thousands or even millions of years for all of our star groups to be so shuffled that no one living today could recognize them. To be sure, Sirius and our other nearby neighbors, then long gone from the solar neighborhood, will appear as nondescript faint stars. Others now approaching us will shine brightly in their places. H. G. Wells realized this when he moved his intrepid time traveler forward on his time machine to the year A.D. 802,701. In one nocturnal episode, when the Time Traveler and Weena are fleeing the threat of the Morlocks, Wells remarks, "Southward (as I judged it) was a very bright red star that was new to me; it was even more splendid than our own green Sirius." Although the reason for it is unclear, some writers of the past report a greenish tint to Sirius and other stars we think of as blue.

The narrator of the tale continues to comment on the firmament of a distant future:

> Above me shone the stars, for the night was very clear. I felt a certain sense of friendly comfort in their twinkling. All the old constellations had gone from the sky; however, that slow movement which is imperceptible in a hundred human lifetimes, had long since rearranged them in unfamiliar groupings. But the Milky Way, it seemed to me, was still the same tattered streamer of star-dust as of yore.

The impression that the steadfast Milky Way arches amongst stars in unfamiliar patterns is also correct—almost. Some simulacrum of today's star-figures will still exist in that distant epoch. Above all Orion, the central one, will in that far-off time still bear a likeness to his present self, as we know from the astrometry of position and proper motion, for all but one of its bright stars move nearly together in a giant stream. To be sure, his shoulders will be grotesquely broader, since Betelgeuse, the odd man out but still distant, will have shifted about six degrees to the northeast, and Bellatrix, the star representing the other shoulder, will be seen about two degrees to the southwest. The other bright stars will appear nearly as they do now. The solar system will have traveled 50 light-years in that time, not one-twentieth the distance to those stars. Perhaps more than any other asterism, this luminous figure made its mark on our hominid past, and will continue to do so in the sky of the distant future.

What an irony it is, then, that this most indelible of star-figures should be composed of the most ephemeral among stars. We noted earlier that Orion is made up of very young stars and that star formation is still proceeding in the very visible nebula in the sword, below his belt. At their distance of some 1400 light-years, the bright stars must all be very luminous indeed. They are several thousand times, and Rigel and Betelgeuse are tens of thousands of times, the Sun's brilliance. But they are only 10 or 15 times as massive as our Sun; hence, they exhaust their fuel at a most unseemly rate. Then after a thrilling finish to their hour upon the stage, they are gone.

The Sun has been itself for almost five billion years and will endure for at least another five billion before it comes to the end

Orion today (left) and as it will appear in the year A.D. *802,701 (right). Betelgeuse and also Rigel may not exist as visible stars at the later epoch.*

of its lifetime. But Rigel and Betelgeuse may live for maybe a million years at most, hardly an instant in star time. And we know that it and Betelgeuse are now in the final stages of their lifetimes. Betelgeuse is already terminal, with only thousands of years left to it. Sometime in the next ten thousand years this red supergiant will burst forth as a supernova and rival the Moon in brilliance. Then in a few years, it will fade away to a stellar cinder. The second-magnitude stars of the belt will last longer, but not much. Orion may recognizably mark our skies longer than any other asterism, but it could just end in a unique fashion, astrophysically rather than astrometrically, with its individual stars passing into the stellar graveyard before it loses that memorable form.

The fastest-moving star known to us is a faint little red dwarf star, just six light-years away. It is known as Barnard's Star, named after the astronomer who first called attention to its rapid motion in 1916. It is the second closest to us, after Alpha Centauri. Yet it shines at magnitude 9.5, too faint to be seen even with binoculars. It sails across the sky at just over ten seconds of arc per year, and covers a degree in under four centuries. This is breakneck speed for a star, but 10,000 years from now, when it passes closest to us, it will move $2\frac{1}{2}$ times as fast at it does at present, and will cover a degree in less than 150 years. Still, at only three light-years from us, it will appear only at magnitude 8, just visible in binoculars. Like the even fainter Alpha Centauri C, known as Proxima, Barnard's star is one of the myriad of little feeble stars that dominate the Milky Way in mass if not in luminosity.

Our faint, inconspicuous neighbor has the distinction of having the most precisely known distance of any object in the universe beyond the solar system. Distances within the system are known to one part in 100 million; thus astronomers have determined the distance to the Sun with an error of something like one mile. But once we step outside, the distances to only a few stars are known to be one part in one hundred, and many of the best known stars, to only a part in ten. The distances to some among even the brightest stars, such as Deneb, Betelgeuse, and Canopus, have been known until this decade only to within 10 or 20 percent, as have all the nebulae and galaxies beyond. It is then small wonder that the size, shape, and fate of our universe is still a matter of some conjecture. As the twentieth century wore on, astronomy, like many other sciences, shifted from an emphasis on observation to one on theory. In some cases we are left with cosmological theories of the 1980s and 1990s, based on fundamental astronomical data of the 1920s and 1930s.

This is changing. In the last few years, European astronomers have taken a giant step toward a wholesale improvement in the observational stellar data. In 1989, the European Space Agency, a consortium of scientific personnel and resources from most of the nations of Western Europe, launched a satellite whose purpose was to make highly precise observations of the luminosities, positions, distances, and motions of all of the 40,000 brightest stars and twice that many fainter ones. Despite a flawed launch, the satellite accomplished its mission. It was named Hipparcos, an acronym, but also in honor of the great Hellenistic astronomer Hipparchus, the discoverer of precession and maker of the first star catalog with magnitudes as well as positions. By now it has produced a veritable mountain of new precise and valuable information about our stellar neighbors. Not as large or spectacular as NASA's Hubble Space Telescope, it may yet prove equal to it in providing the foundation on which a new model of the cosmos would rest. In point of fact, the European Hipparcos and the American Hubble complement each other magnificently, and both are still being supplemented by ground-based astronomical observations with telescopes large and small.

CHAPTER 24

How Well Do We Know How Far They Really Are?

If you haven't measured something, you don't know very much about it.

—KARL PEARSON
(attrib.)

Whenever information is presented in terms of its mean (or average) alone, the manner in which individuals are scattered or spread about that mean cannot be inferred.

Suppose the annual incomes of two groups of wage earners have the same average. One group consists of employees of a single corporation engaged in the same kind of work; the other comprise members of a random audience attending a movie, play, or concert. Could we really expect the two groups to be equally well off financially? Consider the first group. Even if these people aren't members of a labor union their wages aren't likely to differ by much. Differences in seniority, length of employment, frequency of merit raises, and such will lead to a variance within their number, but not a large one.

The second group might easily include some very wealthy people, some not well off, and some in between. Wouldn't one expect these people to be distributed across a much wider range of income? And yet data, at least as presented in the media, may not include such information and may therefore be of little sociological value. Scientific and technical information, however, is rarely considered complete or adequate without an indication of the spread, or dispersion, among the individuals in each group.

Weather information provides another example. Temperatures given for a particular day customarily include the high and the low values reached on that day, along with the average high and low and the record high and low temperatures ever recorded on that date in the past. But those latter figures contain no useful information. To state the highest-ever temperature is comparable to citing the income of the single wealthiest individual in one of the groups of the previous example.

Travel books do no better; again averages are stated but little else. Many of us want to know just how off-average the weather might be during our projected vacation. The fact that the temperature reached 104 degrees one day back in August 1915 tells us nothing about the likelihood of it reaching 80 or 90 degrees in August this year.

By their very nature, data are distributed along a curve, as shown in the figure on page 223. This is the familiar bell-shaped curve, called a Gaussian distribution, after Carl Friedrich Gauss (1777–1855), one of the preeminent mathematicians of all time. Any large number of data points, with few exceptions, will be distributed in or nearly in such a form. The mean is indicated; this point is also the median, or 50th percentile, as well as the mode, the most frequently occurring value and thus the highest point on the curve.

The spread about the mean is called by several names: the standard deviation or, if describing an error of measurement, the mean error or standard error. Its definition is mathematical and unambiguous. From this parameter, along with the mean, the height of the curve and the percentage of the total area under it to the left or right of the mean can be found for any point or value along the way. The properties of the curve are such that just over two-thirds of all cases lie within one standard deviation of the mean on both sides of it. In our temperature example, a standard deviation of three degrees, for example, informs us that on two-thirds of these days in August, the high temperature will be

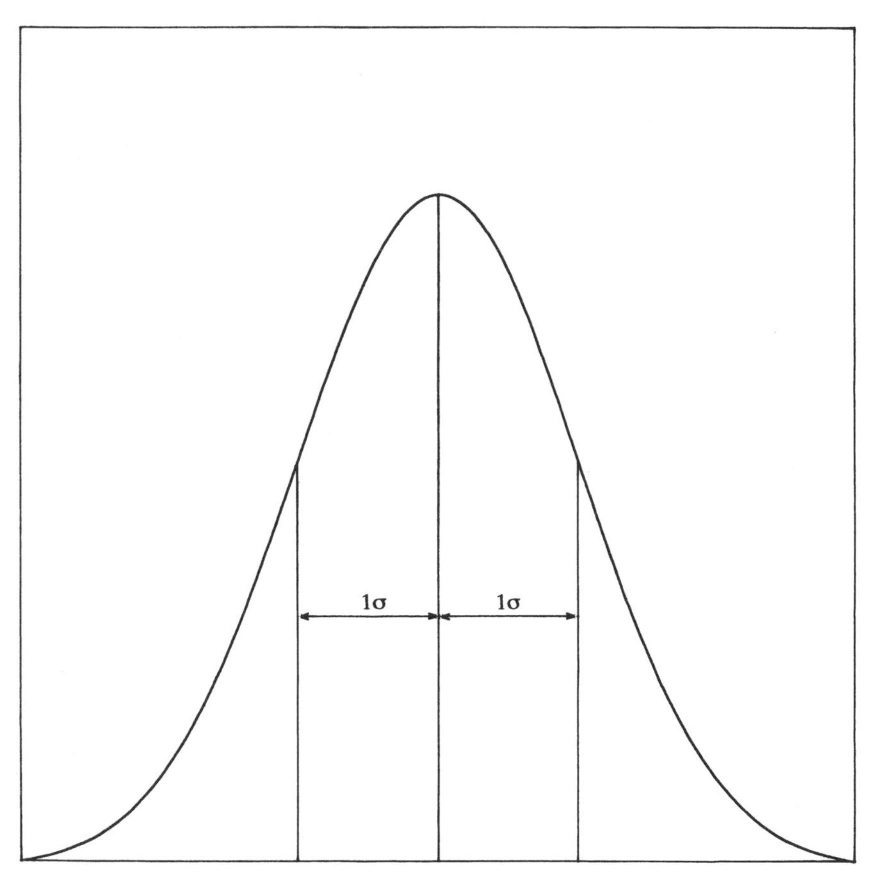

The Gaussian distribution. The mean is shown by the vertical line at the center, and the standard deviation is shown to either side.

within that amount of the mean high for the date. Half of the remaining third, or one-sixth, of the time, the high will exceed the mean by more than three degrees, and the final one-sixth will be over three degrees cooler.

One very clear example of the workings of the Gaussian distribution is found in the I.Q. test. Although lately fallen out of favor, it measures something with precision. A value of 100 is adopted for the mean (presumed to represent the intelligence age divided by the chronological age times 100). On one of the major

I.Q. tests, the standard deviation is 13; thus, within 13 points of 100, or from 87 to 113, are found just over two-thirds of the population. That leaves the remaining third equally divided so that one person in six is found to score above 113 and the same number below 87. Scientists speak of the mean as being 100 ± 13 (100 "plus or minus" 13 points).

This Gaussian distribution also describes the amount of error in a group of repeated measurements. Thus we have found the mean or average distance from the center of the Earth to that of the Sun (the astronomical unit) to be 149,597,870 ± 1 kilometer, or about 92,955,807 ± 1 mile (the error of 1 is approximate in this case). Clearly this most important of all astronomical distances is known to an extremely high level of precision—one part in a hundred million, no less. Other distances within the solar system are known in terms of this one, and are thus of superb quality. NASA couldn't place a space probe just next to Triton, Neptune's big satellite, as it did unless distances were known to this precision. (Precision is not to be confused with accuracy; a clock reading 30 minutes fast is precise to the minute, but it remains inaccurate until corrected.)

The distance to Barnard's Star is the most precisely known distance to anything outside the solar system. The error is 1 part in 420, or 0.24 percent. The distance turns out to be 378,050 ± 900 astronomical units, or 5.975 ± 0.0142 light-years, the error amounting to 5.2 light-days. The distances to a few dozen other nearby stars are known to plus or minus one percent or less. But much beyond a few score light-years, large uncertainties are unavoidable; these distances are just too large for anyone to get a good handle on them, at least at present.

As mentioned earlier, nothing in astronomy needs to be known to high precision more than the distance from the Earth to the Sun. It sets the scale of the solar system and all in it, and it is also the foundation affecting the distance scale for the entire universe and everything in it. The heliocentric or trigonometric parallax, the shift in the position of a star, as illustrated in Chapter 22, uses this distance for its baseline of triangulation. If this astronomical unit were found to be ten percent larger than we now think, the whole universe would grow apace. The mass–

luminosity relationship for stars, the rate of expansion of the universe, and even the speed of light would need recalibration. This is not going to happen; it is no accident that its precision is known so well, when we consider the immense effort that has been involved in its determination. The same uncertainty in the distance between New York and London would amount to two inches.

The Hellenistic astronomers, including Ptolemy, had found the distance to be about 4,500,000 miles, and Kepler tripled this number in the early seventeenth century. Since 1700, it has been set at between 81 and 96 million miles; thus, for three centuries the proper order of magnitude has been established, and the sizes and masses of the Sun and planets known, along with a general idea of the distances to the stars.

Chapter 25

Motions, Then and Now

I know that I am mortal by nature, and ephemeral; but when I trace, at my pleasure, the windings to and fro of the heavenly bodies I no longer touch earth with my feet: I stand in the presence of Zeus, himself, and take my fill of ambrosia.

—Ptolemy

Nothing could be more obvious than that the earth is stable and unmoving, and that we are in the center of the universe. Modern Western science takes its beginning from the denial of this common sense axiom.

—Daniel J. Boorstin
The Discoverers

Every visible object in the sky, except the Sun and the Moon and a few faint extended light sources like the Orion nebula and an occasional comet, appears as a point of light to the naked eye. Before the invention of the telescope astronomers could only record and study the positions of heavenly bodies and, from changes in their positions with time, their motions. With our noted exceptions of the Sun, Moon, and planets, no starlike thing could be seen to move in the sky prior to the invention of the telescope, except in lockstep with all of the other stars. Moving together they reflected the rotation and revolution of the Earth and the precessional motion.

The simplest motion to observe and understand is that of the rotation of the sky full of stars, the sidereal motion, and it illustrates best the conceptual model of the celestial sphere. This is, of course, an incorrect model, but it is a perfectly satisfactory one even today for some purposes such as surveying and navigation. As explained earlier, the Sun's motion is the next simplest to comprehend, followed by that of the Moon. Finally, the planets show motions of the greatest complexity.

Since motions were about the only subject of study in the days of pretelescopic astronomy, astronomers dealt exclusively with cosmology, which was defined in ancient and medieval times as the understanding of the layout of the solar system with the stars simply tacked on at its edge. No study of the physics of the planets and stars was possible until long after the invention of the telescope.

Any successful cosmology must be required to explain a number of observed phenomena. Among them are the sidereal motion of the stars reflecting the rotation of the Earth, the annual motion of the Sun along the ecliptic at an angle inclined to the direction of rotational motion, the motion and the phases of the Moon as they change throughout each lunation, and the revolution of all of the planets about the Sun in the same direction, as well as the fact that they lie nearly in the same plane. Seen edge on, the orbits of the planets would appear as shown in the figure on this page.

Various schemes were successful to different degrees in explaining these observed phenomena. Other phenomena not mentioned earlier, but still noticeable with the naked eye, are of

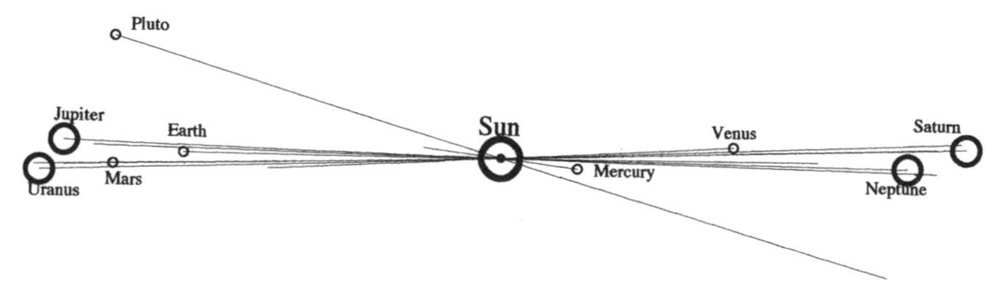

The solar system as it would appear if seen edge on. The Sun and planets are not drawn to scale.

greater complexity and require more time to observe, such as the tides, the frequency and predictability of eclipses and their cycles, and of course the very slow precessional motion.

The cosmological models developed by most civilizations were not sufficiently grounded in these kinds of observational details and fell far short of explaining them. The majority of them were content with the development of a calendar to keep track of the seasons and other records, and the Babylonians, Mayans, and some eastern Asian societies developed more accurate calendars than our own. Some of the more sophisticated ones could handle eclipse predictions as well. These activities were almost always the dominion of a favored few, a priesthood who zealously kept their arcane knowledge to themselves in order to retain their rank and influence. The schemes in other places, sometimes of great elaboration, that were conceived to explain the sky fulfilled the mythic needs of particular cultures while showing little regard for the observations they made, even though some early societies, as already noted, developed calendars of greater precision than our own.

Then, sometime around the sixth century B.C., on a rocky peninsula stretching like a skeletal hand down into the Mediterranean Sea from the Balkan Peninsula, a sea change occurred in human thought. Perhaps the arrival in Greece about 750 B.C. of an alphabet from the Phoenicians and others—it's not easy to handle nuance and metaphor in hieroglyphic scripts—led to concepts not otherwise easily communicable. In any case, the Ionian astronomers (Thales, Pythagoras, Anaxagoras, and others) came up with an understanding of lunar phases and the causes of eclipses, and conceptual astronomy began.

From the Ionians of the sixth century through the Athenians led by Plato, Aristotle, and Eudoxus, the Greek scientists showed a desire for building a model of the cosmos based on observation. At the outset, the early Ionians had a grasp of some things. Their Earth was spherical, not flat, and they recognized that the Moon and Sun were also round, not flat or disklike. They knew how eclipses and the lunar phases worked as well.

These same people developed geography at this same time. The Eastern Hemisphere, at least, began to resemble its true outline, and from this point through the time of Ptolemy, much of the world they knew was mapped out.

The concept of proof came in then or soon afterward. Aristotle proved to his satisfaction the sphericity of the Earth by noting that the edge of its shadow on the Moon during a lunar eclipse was always an arc, a segment of a circle. Only a sphere casts a circular shadow in every direction. This test still constitutes one of at least four proofs for the Earth's sphericity; some of the others were also known at that time. In Chapter 10, we found that the angle of the north star in the sky is equal to the latitude of the observer. Only a ball can maintain this relationship at all longitudes. Furthermore, a ship observed to head out to sea is seen from the harbor to appear to sink into the ocean. In truth it is sailing over the curved lip of the horizon, and the effect is noted from any point on the globe. One more proof concerns the dip of the horizon. We can see this in the figure on this page. Since the top of the spectator's head is at a level with my eyes and camera

Atop the John Hancock Building in Boston, the dip of the Earth's horizon is evident. If the Earth were flat, the horizon would be seen passing just above the person in the photo.

lens, the horizon line, if the Earth were flat, should be seen to intersect her image at that point. It does not; it appears half a degree lower because we are about 800 feet above its surface, and the Earth no longer fills the full lower hemisphere. It takes only a modest height like this to show the so-called dip of the horizon, in this case about 35 miles away.

Developed during the second and third centuries before Christ in and around the great Egyptian port city of Alexandria near the Nile Delta, the cosmological models of the later Greek scientists formed one of the two greatest astronomical developments of antiquity. For the first time, a quantification of the size of the universe was achieved through rational thought. These people realized that the Sun was not a smallish object, but a sphere several times as large as the Earth.

The other great accomplishment consisted of the superb observations recorded by the Babylonians during the first millennium B.C., the most precise observations in the world prior to the Renaissance. The Greeks were not as committed to or skillful at observation, but they provided a real emphasis on "saving the phenomena," that is, making theory fit the observed data. Their discoveries in the eight centuries between 600 B.C. and A.D. 200 were truly a great achievement individually and collectively. The fundamental difference between the later Greeks and Babylonian astronomy was that the Babylonians sought to predict only a few specific phenomena. But Greek cosmology had as its goal the determination of planetary positions at any time of the past, present, or future. The ability to predict is now, more than ever, accepted in modern science as a defining hallmark of the merit of a theory.

The extraordinarily sophisticated Greek model that contributed so much to the legacy of the Middle Ages was the system of Ptolemy (Claudius Ptolemaeus, ca. A.D. 100–170), which was built on features of the system of Aristotle (384–322 B.C.) and the foundations laid down by Ptolemy's three most famous Alexandrian predecessors: Aristarchus (ca. 310–230 B.C.), Eratosthenes (ca. 276–195 B.C.), and Hipparchus (ca. 190–120 B.C.)

Ptolemy described the scheme in his book, known now by its Arabic name, the *Almagest*. It is more of a textbook than an account of discovery, and it dominated astronomical thought for over a millennium, even providing a model for Copernicus (1473–1543) when he published his alternate theory just before his death in 1543.

Whereas we are privy to a lively correspondence Copernicus maintained with several other astronomers as he developed his cosmology, we know so little about Ptolemy that it is impossible to establish the extent of his contact with contemporaries. The fact that he lived three or more centuries after his illustrious predecessors is noteworthy. He must have had access to the great Alexandrian library and he made many observations himself, which he combined with earlier material. From this distance he appears, like Bede, to stand alone in his time.

A widely known illustration purporting to be a sixteenth-century woodcut reveals the growing curiosity about the universe. Actually the drawing was made in the nineteenth century.

The cosmological design of Ptolemy was geocentric, and because it is based on this one glaring misconception, his illustrious achievement was destined to be depreciated. Ptolemy's masterpiece, however, is worthy of the fullest admiration, for its consistency and logic are of the highest caliber. The direct confrontation between the *Almagest* and the alternate scheme developed by Copernicus touched off, as much as anything, the intellectual fervor that led to the Renaissance, the Reformation, and the abandonment of the medieval tradition. A well-known woodcut (reproduced in the figure on page 232) captures the spirit of discovery of that time, even though it is of much later origin.

Not long after the Roman Empire gave way to the early Middle Ages in Western Europe, Islamic scholars in the Middle East became active. They added little conceptually to the existing

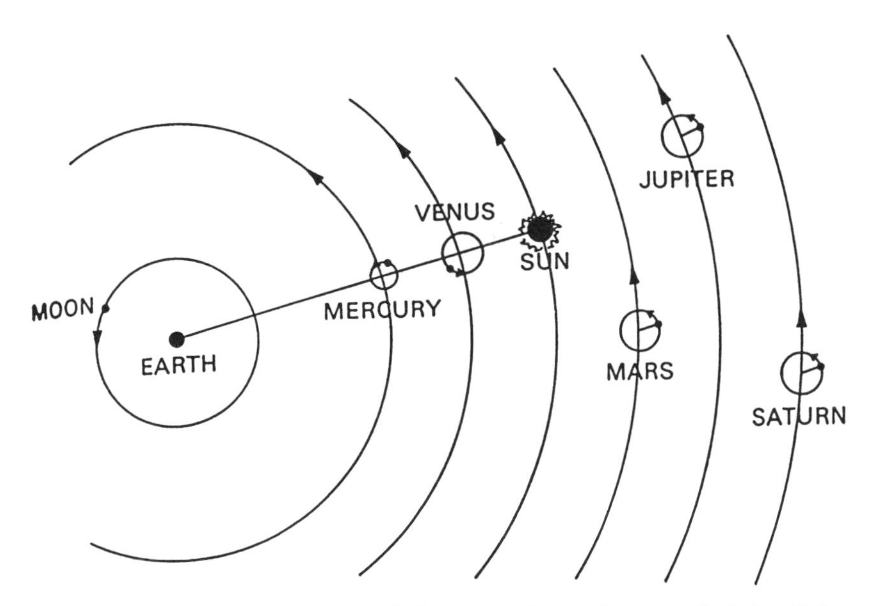

The geocentric system of Ptolemy. Each planet moved around on a small circle called an epicycle, which orbited on a larger one called a deferent. The periods were determined in such a way that they reproduced their observed positions very closely. S. P. Wyatt and J. B. Kaler, Principles of Astronomy: A Short Version, *2nd Edition. Reprinted with permission from James B. Kaler.*

Ptolemaic model, but they continued to fine-tune and improve it partly through new observations.

The medieval years were not completely devoid of scientific activity in Europe. But they did witness an abandonment of most of the complexity and quantitative sophistication of the Ptolemaic model. A comparison of Ptolemy's cosmos with the very simplified version of it portrayed by Dante in his *Divine Comedy* reveals a great deterioration in the comprehension of its details and the ability or desire to save the phenomena, the cornerstone of Greek astronomy. As we see in the figure on page 235, Ptolemy's model had been denigrated and corrupted by Dante's time into an attempt to depict the locations of Heaven and Hell and little else.

Such learning as there was then centered on commentaries of doubtful quality written by contemporary scholars and reverting, in astronomy, from Ptolemy to the simpler, more physical, and less mathematical world of Aristotle. Slowly over this period, with the encouragement of the Church, learning became more doctrinaire. Formal learning was divided into two distinct parts called the trivium and the quadrivium. The trivium was the basic form of education, and for centuries it constituted the only education most literate people ever received. As its name implies, it consisted of three subjects: grammar, rhetoric, and dialectic (logic). The quadrivium was based on a more mathematical structure and consisted of arithmetic, geometry, music, and astronomy. Its strong mathematical basis led to its demise, due in part to the increasing shortage of people who could teach one or more of its disciplines.

As a result, education became more a matter of faith and dogma, replacing the empiricism of observation and experiment. Astronomy retained the concept of a spherical, stationary Earth surrounded by the shells or realms for each of the seven moving bodies with one for the stars. All motion was held to be circular, and the dualistic material nature (Aristotle's four elements—earth, air, fire, and water—for the Earth and a fifth substance, the quintessence of purity, for the heavens) still held sway. Furthermore, the heavens remained changeless (aside from the well-observed motions) and immutable. This age took its cue from

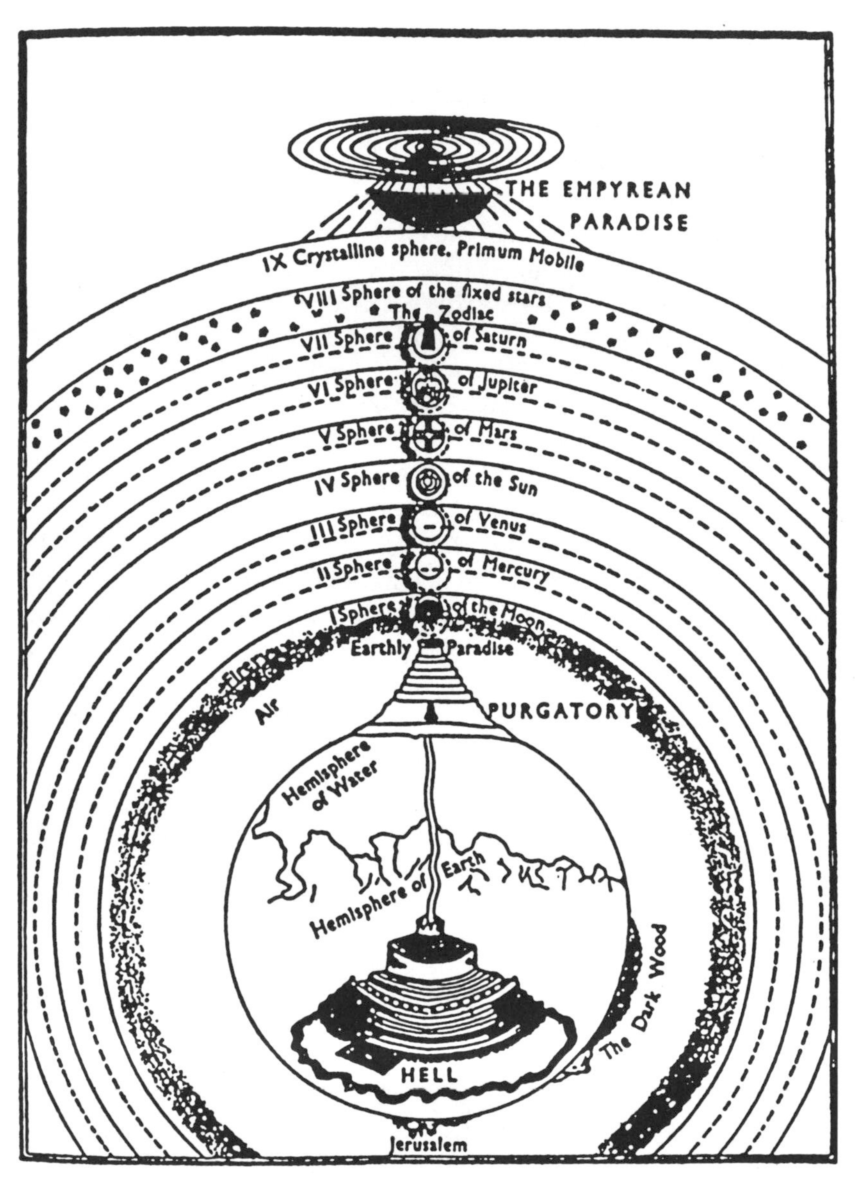

The universe as presented by Dante in the Divine Comedy. *Notice the emphasis on ecclesiastical locations not involved in Ptolemy's cosmos.*

writers like Saint Augustine (354–430). In Book V of *The Confessions* he maintained that "one who can measure the heavens, number the stars and balance the elements is no more pleasing to God than one who cannot, and that scientific knowledge is more likely to encourage pride than to lead people to God." Salvation was the goal, not material progress; science was not only superfluous to that aim, but might even be dangerous.

Astronomy came to feel the effects of the Church, as it maintained that the Ptolemaic system was at best a convenience for calculations but not necessarily real, whereas the physical and metaphysical substance of Aristotle was closer to reality. Explanations were dependent on what God wanted to do and did rather than in terms of cause and effect.

By the thirteenth century the first universities had arisen, and translations of Aristotle and other ancients became available in Latin, many for the first time. Only then was Aristotle's world opened up in full to medieval scholars. As they read his works, they began to question them. Scientists of the time blended the best of Greek theory with Babylonian observation, to which they also gained access. Some proposed that the Earth rotated on its axis. If so, this meant that our small central sphere need rotate only at a modest rate, whereas the alternative had the entire heavens in rotation, with the uppermost spheres containing the stars and God's abode rotating at a fearful and unseemly speed. From then to the time of Copernicus, a number of people searched for new and better ways to explain the cosmos.

After its publication in 1543, Copernicus's world system competed with that of Ptolemy, with the outcome very much in doubt. This period was one of much astronomical activity by people who defended one or the other system until Tycho Brahe (1546–1601) proposed a model that was a compromise, political as well as astronomical. In it, the planets circled about the Sun, which then orbited the stationary Earth carrying them along as satellites. From Venus and Mars, our Earth would appear to the naked eye as a brilliant blue star and at times, first to one side and then some two weeks later to the other, the Moon could be seen as a fainter but still bright white star. If either Venus or Mars had

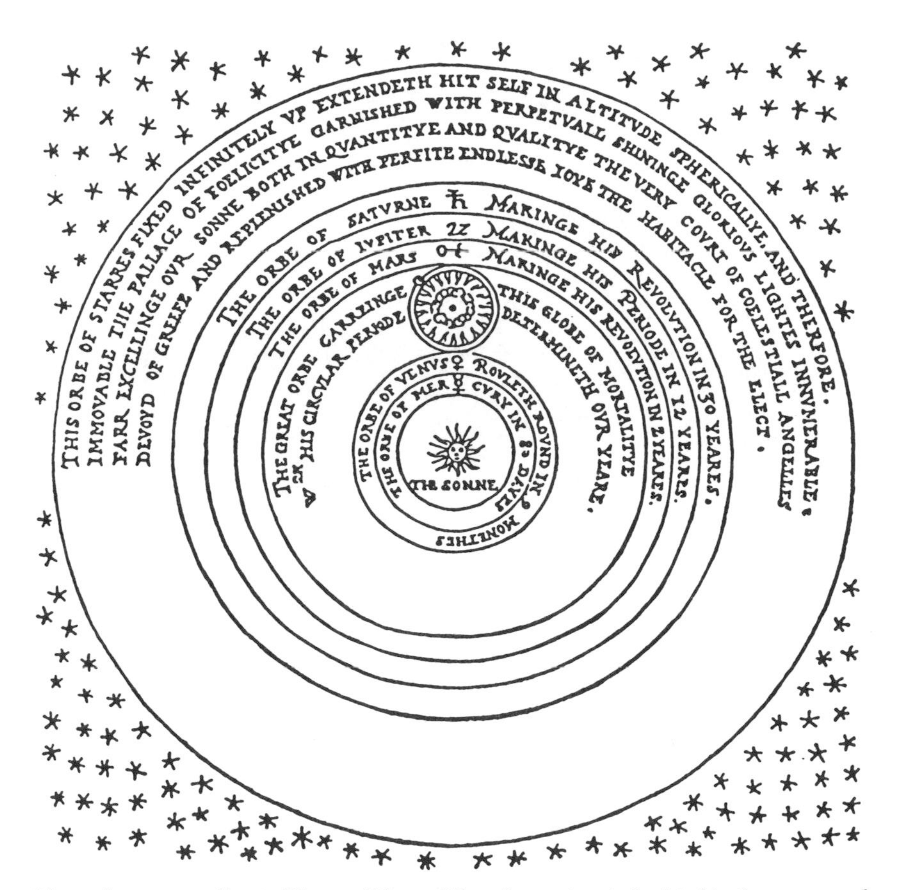

The universe according to Thomas Digges. The solar system is depicted in the manner of Copernicus, but the stars are shown at different distances well beyond the orbit of Saturn. Digges realized this implication of the heliocentric theory well before Galileo and Kepler. From Digges' Perfect Description of the Celestial Orbes, *published in London, 1576.*

such a large companion, we too would see a double planet in our sky. What would have been the effect of such a sight on the cosmos of Aristotle or Ptolemy? Might the Copernican model have been established earlier or with more certainty? Of course we can never know for certain, but the geocentrism that pervaded more than fifteen centuries of thought and custom might

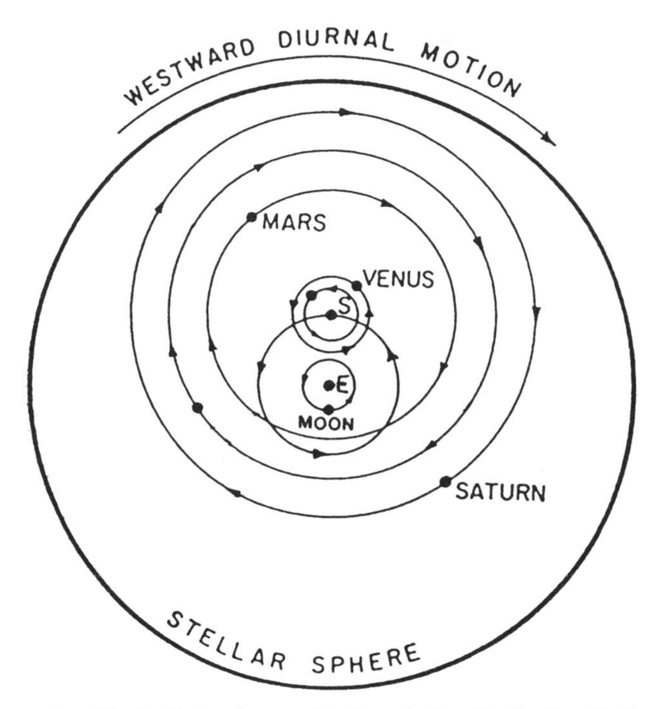

The compromise of Tycho Brahe showing the Sun circling the Earth with the planets as satellites. From De Mundi, *by Tycho Brahe, published 1588.*

have eroded more quickly. Speculation of this kind can provide many insights into our understanding of the solar system and how we came to understand its true nature.

We tend to think of the next step as belonging only to the two great contemporaries: Galileo Galilei (1564–1642) and Johannes Kepler (1571–1630), whose system of elliptical orbits is illustrated in the figure on page 239, the first to show them to the proper scale. But several important developments occurred in between. Among them was the nova, or new star, appearing brilliantly near Cassiopeia in the northern sky in 1572 and known as Tycho's nova. How could the sky be immutable if a new star appeared? This and a growing awareness of the greater distances to the stars by Thomas Digges (ca. 1546–1595), Giordano Bruno (1548–1600),

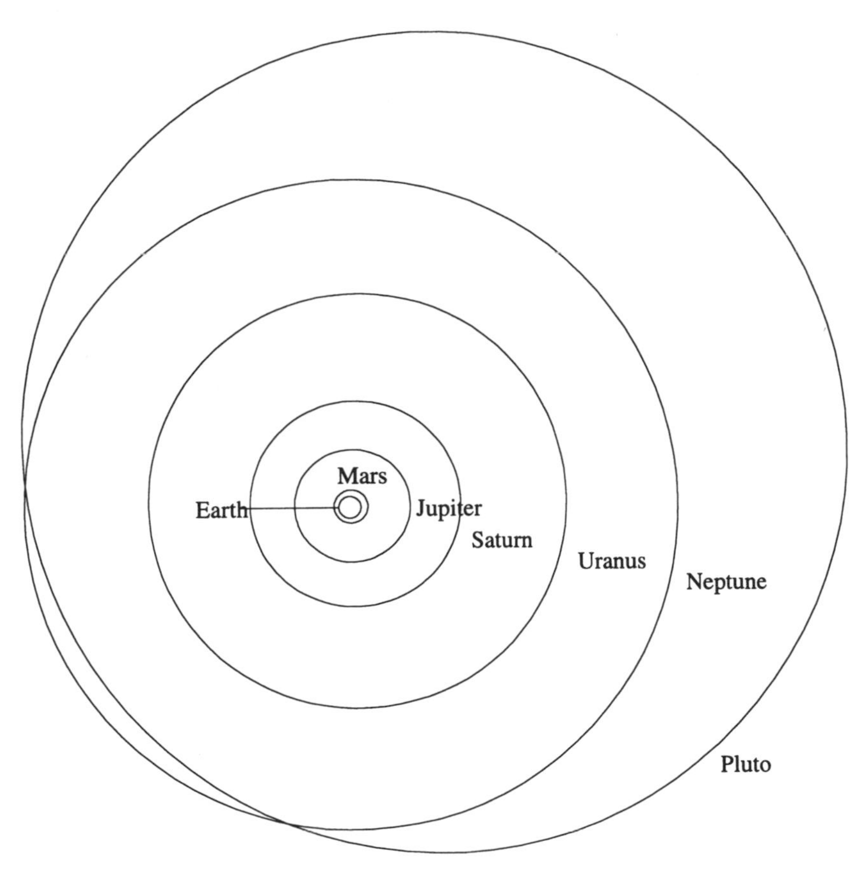

The system of Johannes Kepler was the first to show the planetary orbits as ellipses and to reveal their correct relative sizes. The orbits of Uranus, Neptune, and Pluto are shown, although they were unknown to Kepler. The orbits appear circular at first glance because their eccentricity is small.

and many others in the latter half of the sixteenth century led to a better realization of the observations necessary to confirm or abandon either system.

Somewhere during this debate, the concept of the scientific method came into being, bringing with it probably the greatest revolution in history. Albert Einstein, while discussing the state of non-Western science, remarked that

> the development of Western science is based on two great achievements, the invention of the formal logical system [as in Euclidian geometry] by the Greek philosophers, and the discovery of the possibility to find out causal relationship by systematic experiment [in the Renaissance]. In my opinion one has not to be astonished that the Chinese sages have not made these steps. The astonishing thing is that these discoveries were made at all.

Not all of the ensuing dialogue over the different schemes was confined to scientific grounds. The heliocentric theories all but do away with the locations, if not the basic Christian concepts, of Heaven and Hell. These and other ecclesiastical concerns were very much involved. Probably no one held Copernicus and his theory in lower esteem than did his contemporary Martin Luther. Luther astutely recognized, as did some other religious figures of the times, the Copernican cosmology not only as a replacement for the Earth-centered Ptolemaic world system, but also as a competing worldview on somewhat equal ground: a schema based on testing and questioning, as opposed to one of blind faith in supreme order. The choice of observation and experiment over dogma was central to the developments that brought about the end of the medieval period in Europe and led to the scientific revolution of the following century. Astronomy, as much as any other science, introduced this revolution.

CHAPTER 26
The Age of Aquarius

The Ram, The Bull, The Heavenly Twins
And next the Crab, the Lion shines,
The Virgin and the Scales,
The Scorpion, Archer and Sea Goat,
The Man who held the Watering Pot
And Fishes with glittering tails.

—ISAAC WATTS

Your horoscope is two thousand years out of date. As we saw in Chapter 9, the astrological arrangement of the signs of the zodiac was keyed to the celestial alignment of classical times and the Roman Empire, as well as for the beginning of the Christian era. Today the vernal equinox point is located not in Aries but in western Pisces, which it is about to leave to cross over into Aquarius, thus initiating the "Age of Aquarius."

Each of the 12 signs corresponds to a constellation, but each has shifted considerably away from its corresponding group of stars. Furthermore, as the constellations are drawn, Ophiuchus has a foot that pushes its way through the zodiac, and this star group actually occupies more of the ecliptic than does Scorpius, for which this segment was named. Cetus, the sea monster, also thrusts his head into the zodiac near Pisces. The signs traditionally form the background for the zodiac and are the framework on which modern astrology rests. Since the solstitial points used to be in the constellations of Cancer and Capricorn, the tropics, the limiting latitudes at which the Sun can be seen overhead at the zenith at least one day of the year, were named for them. These constellations along with Aries and Libra, in which the equinoxes fell, contained the four cardinal points of the zodiac. The table on this page shows the commonly accepted dates during which the Sun was within each of the 12 signs.

But precession changes all that. For one-twelfth of its period of 25,800 years, or 2150 years, the four colures grind slowly backward through each zodiacal sign, until the arbitrary limits between one sign and the next are reached. Sometime near the beginning of the Common Era, the vernal and autumnal equinoxes crossed the lines separating Aries from Pisces, and Libra from Virgo, respectively, and the summer and winter solstices backed from Cancer and Capricorn into Gemini and Sagittarius. Many notations, to be timely, would have changed names accordingly, and two of Henry Miller's best-selling novels, for example, might well be known as the Tropic of Gemini and the Tropic of Sagittarius. Astrology columns would begin with Pisces, the sign

Sun Sign	Dates
Aries	Mar 21–Apr 19
Taurus	Apr 20–May 19
Gemini	May 20–Jun 20
Cancer	Jun 21–Jul 22
Leo	Jul 23–Aug 21
Virgo	Aug 22–Sep 22
Libra	Sep 23–Oct 22
Scorpius	Oct 23–Nov 21
Sagittarius	Nov 22–Dec 21
Capricorn	Dec 22–Jan 20
Aquarius	Jan 21–Feb 19
Pisces	Feb 20–Mar 20

of the fish, and this early Christian symbol would be at the head of the zodiacal procession.

Those passages were made about two thousand years ago. Aren't we now about due for another change of names? We are—hence, the Age of Aquarius. But recall that the locations of the astrological signs as well as the astronomical constellations are both arbitrarily defined. Each sign spans the twelfth part of the circle and is 30 degrees in width, whereas the constellations (14 of them with Ophiuchus and Cetus) vary in width, and both are set without reference to the sky. Furthermore, differences are found between the sign limits of one society and those of another. As a result, the Age of Aquarius may be said to begin any time from A.D. 1781 to 2614, depending on whose definition is adopted, with

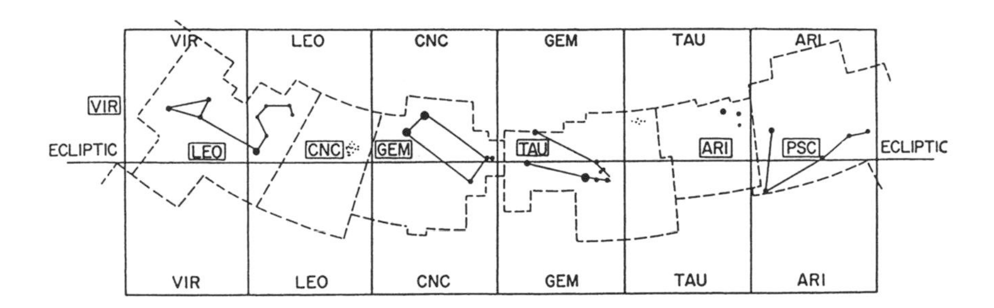

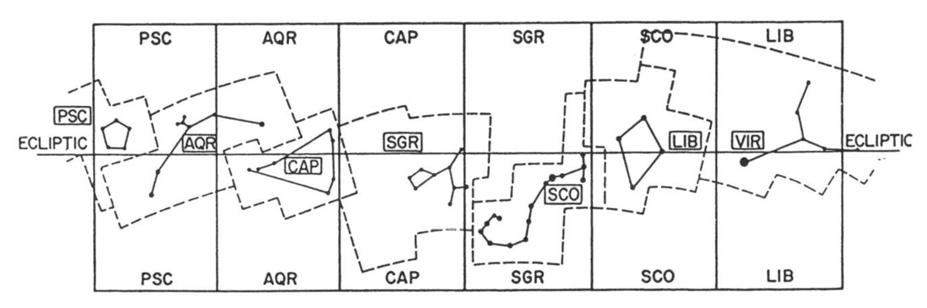

The signs of the Zodiac are shown against the constellations of the same name. The constellations with their irregular boundaries are displaced from their respective signs. The boundaries of both constellations and signs are arbitrarily defined. R. E. Culver and P. A. Ianna, The Gemini Syndrome. *Reprinted with permission from Philip J. Ianna.*

2614 being the best match for current constellation boundaries. Neither set depends in any way on natural phenomena and both exist only for convenience. The delineation of both is depicted in the figure on page 243.

* * *

The period of 2150 years per sign is coincidentally equal to that of a cycle of civilization, according to Oswald Spengler. In his seminal work, *Decline of the West,* he equates turning points within one civilization and another by this amount; thus, for example, Napoleon and Alexander are spaced by this interval, and Spengler considers them to be equivalent figures in classical civilization and in our own. Although Spengler's theory of cycles and recurring events has passed out of favor among historians, it may still enjoy a fashionable currency among certain subcultures in Western society.

The premise on which astrology rests is straightforward. It is that the apparent positions of the Sun and, to a secondary extent, the Moon and planets affect one's personality and life, successes and failures. This code of beliefs probably harks back to the dawn of recorded history, if not earlier. The seven celestial objects that are seen to move against the background of the so-called fixed stars (fixed in the sense that over a single lifetime they appear stationary to the casual observer) have long been imbued with supernatural properties. In some cultures they are gods; in others, the manifestations of gods. But in either case they are sublime and of godlike essence.

Ptolemy was one of the authorities upon whose work Western astrology has been fashioned. While his *Almagest* discusses the cosmos and his intricate model for its interpretation, his *Tetrabiblos,* the "four books," has become a touchstone of present-day astrology.

The early Christian Church was not sympathetic to astrology, although its opposition was not very strict. One of the problems for the Church was the determinism of astrology, which has an impact on questions of free will and individual responsibility; the other is the polytheism inherent in the belief that the planets are gods or the visible manifestations of gods. To a lesser

This woodcut is a representation of the zodiac and shows Ptolemy at the lower right. From Textus de Sphaera, *published in 1531.*

extent, some Church leaders were troubled by the charlatanism in the field as well.

Ptolemy laid down, as much as anyone, the rules of the game. In current astrology, the customary datum is the sun sign, the sign of the zodiac against which the Sun appeared at the moment of one's birth. Thus each of us is a "Taurus" or a "Leo" or a "Pisces," as the case may be, if the Sun was within the limits of that sign as seen from the place and time of our birth according to the dates listed in the table on page 242.

Almost everyone knows his or her sun sign. We may even have coffee mugs or T-shirts proclaiming our birth signs, but this alone cannot be taken for belief in astrology. Numbers vary as to the percentage of adults that do believe or practice this discipline, but in the United States, the figure is not far from 50 percent.

Astrologers who know and practice their trade, however, look beyond the sun sign. The six other movable objects, the Moon and the five bright planets, are also alleged to influence our personalities and our fates. A complete horoscope takes their locations and mutual alignments at the moment of birth into account. Differences are found between the methods of one astrologer and another, but the underlying astronomy is based on a few internally consistent rules.

The foundation is the horoscope. This is a concept, illustrated by a simple diagram, as depicted in the figure on page 247, that divides the sky, particularly the part containing the ecliptic and the zodiac, into 12 equal sections, one for each sign. At the time of one's birth (the time of conception would be more to the point, but who knows when and where that took place), the horizon divided the sky into two equal shares. Somewhere the zodiacal region is cut by the eastern and western horizons, labeled the ascendant and descendant, respectively. The first 30-degree segment of the ecliptic extending downward from the eastern horizon forms the first "house," the next segment, the second house, and so on. Thus the segments just below and above the western horizon form the sixth and seventh houses, respectively, and the twelfth house is the segment just above the ascendant. The 12 houses may coincidentally agree with the 12 signs, but at any moment they are generally offset from them. The

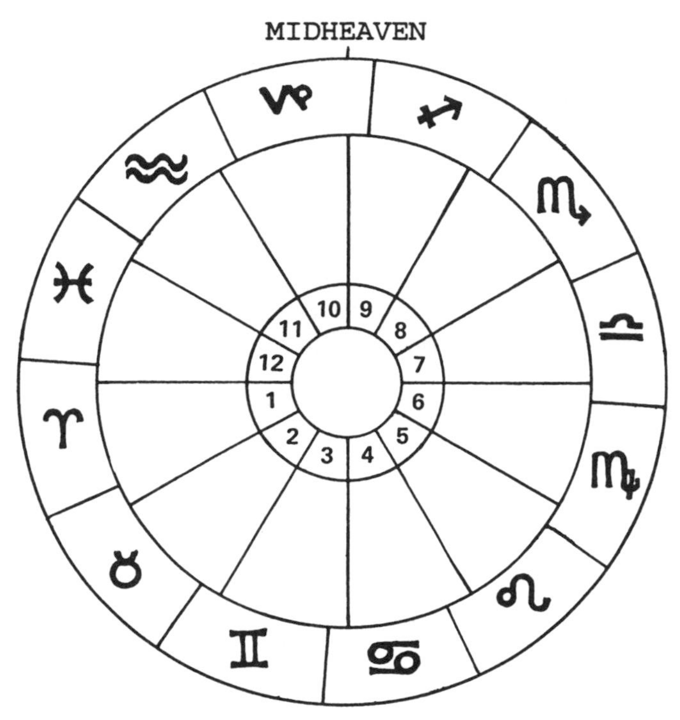

A simple horoscope with the 12 signs and 12 houses. Midheaven is the point due south where a planet appears highest in the sky. The ascendant, or eastern horizon, lies and forms the boundary between houses 1 and 12, and the descendant, or western horizon, does the same between houses 6 and 7.

date fixes the Sun, Moon, and planets in and among the signs, and the time of day fixes them within the houses as well. The arrangement for your moment and place of birth is your horoscope.

Western astrology, which differs from that practiced in the Eastern world, can be divided into two types: natal and horary. Natal astrology presumes that the configuration of the planets at birth affects character and personality. Horary astrology deals with the immediate future of individuals (a third type, mundane astrology, does this for institutions and nations). Former President Reagan's favorable days—days on which major decisions were best made—were based on predictions of this type. The sun-sign variant comprises the mainstay of newspaper columns in which the position of the Sun alone figures in the forecast.

Many factors contribute to the negative opinion that scientists have of astrology, but predominant is the feeling that the discipline does not allow itself and its hypotheses to be subjected to critical and unbiased investigation. Science is a self-correcting affair, probably the most self-correcting and self-critical human activity of all. In that context, certain forces have been deduced that astrology chooses to ignore. Gravitation has been found to operate only along the radial direction from a planet to the Earth, whereas astrology considers only the transverse direction in the plane of the sky to be of influence. And why don't most astrologers accede to precession and place the Sun in the right star group at the right time? Since astrology was codified, we have discovered three planets whose influence should surely be taken into account. But how can tiny distant Pluto influence us as much as nearby Venus or mighty Jupiter? Isn't the Earth of influence and shouldn't it be included in horoscopes of astronauts in space or on the Moon? Many questions arise for which there is never an answer. Astrology is placed among the pseudosciences for this reason; no modification to its doctrine has been made in 20 centuries.

Impartial tests of the efficacy of astrology are available. If the names of the signs were deleted from the daily astrology column, shouldn't people be able to pick out the information that applied to them after the fact? Tests have shown that they do so no more often than one time in twelve, as guesswork would score. What about two constellations, Cetus and Ophiuchus, that actually contain part of the zodiac; why aren't they in the lineup? On the first Tuesday after the first Monday in November, a major event takes place every alternate year, when Americans go to the polls. One candidate for each national office will have a good day and one a bad one; this much we know in advance. On the preceding day, however, astrologers usually decline to predict the outcome of an election.

People will continue to believe what they want to believe, regardless of fact. *Mumpsimus*, defined as a continued belief and adherence to exposed but customary error, or a custom or tenet proven wrong, is just as prevalent today as it was in medieval times. Most people still believe that crime and aberrant behavior increase at the time of the full Moon, despite incontrovertible evidence to the contrary. And who among us does not believe that snowfalls were more frequent and greater when we were young?

Chapter 27

Stonehenge and a Tower in Rhode Island

In Touro Park near the highest point in the city of Newport, Rhode Island, stands a small stone tower. I first saw it as a child, and I was told at the time that no one knew when and by whom it was built or for what purpose. The local people contended that it was built in early colonial times to serve as a mill, but belief in a much earlier Scandinavian origin had never completely disappeared. The means now appear to be at hand to settle this perplexing controversy, and the result may reshape our ideas of the early settlement of this continent by people of the North. In the course of the study, the nascent science of archaeo-astronomy will again demonstrate its power to date ancient structures and, at times, alter history. At issue may be the question of the oldest European structure in America.

The far northern Atlantic Ocean has a geography all its own. Unlike the other great oceans with vast unbroken expanses of water containing at most only a tiny island or two far from any continental landmass, it has a series of giant islands spread like stepping-stones from Europe to North America, none separated by more than a few hundred miles from its neighbor. In earlier times whenever climatic conditions permitted, travel across the northernmost Atlantic waters could have been quite a bit easier than across the ocean at other latitudes.

The tower stands in Touro Park near the highest point in Newport, Rhode Island.

Today our view of the Atlantic Ocean is one of three thousand miles of unbroken water. It is crossed at a single long stretch by jet aircraft or by ship, as it was in 1492 by vessels under the command of Christopher Columbus. Anyone whose transatlantic flight has been diverted northward, over the northern wastelands, knows them to be inhospitable to any but the sketchiest human existence. It is not easy for us to realize that even in historic times this was not always so, because the climate of the Earth is not steady but undergoes swings in temperature, the largest of which caused the great ice ages when sheets of ice covered much of Europe and the United States and Canada. The first settlement of the Western Hemisphere occurred over 10,000 years ago, when Asian people migrated here from eastern Siberia. As the ice age was coming to an end, glaciation was still so extensive that much of the world's sea water remained in frozen form, ocean levels were lower than they are at present, and the Bering Strait formed a dry land bridge between the continents of Asia and North America. Ever since the end of this most recent ice age our climate has vacillated between periods of relative warmth and coolness, often lasting for centuries.

One of the recent warm periods began with the rise of the Greek city-states and the Roman Empire and lasted well into the late Middle Ages. During this period the lands of the North Atlantic flourished to a degree beyond anything seen since. It was toward the end of this climatic optimum that they had their finest hour upon the stage of human affairs. Sometime toward the end of the first millennium A.D., the Nordic races of Scandinavia developed their amazing longboats, those sleek, speedy craft that allowed the Vikings to raid and pillage many of their European neighbors to the south and west. About the year 1000, this barbaric activity tapered off and a period of exploration and settlement began. The Norsemen discovered and settled southern Greenland. Agriculture there flourished to the extent that a number of communities arose along the coasts. By about 1200 Greenland even had its own bishopric with 17 churches and several monasteries, and seafaring and trade were extensive.

Then in the fourteenth century, something happened to the weather. For reasons still not fully understood, the climate of Greenland cooled markedly. The Gulf Stream was diverted to the south and Greenland and Iceland were denied its warm benevolence. The harbors of the villages became icebound and trade and agriculture became impossible. Contact with Europe was severed and the European inhabitants of Greenland vanished in a manner yet unknown. Possibly they intermarried with the natives living there, or maybe they simply froze or starved to death. In any event Greenland and its settlements became lost and forgotten. We now know this largest of all islands to be a key to the climate of Europe, because about a century later northwestern Europe also cooled and winters became much harsher. This period, known as the "Little Ice Age," lasted until the middle of the nineteenth century when European weather turned warmer again. But Greenland remains mostly a frozen wasteland to this day.

How far did the successors to the feared Vikings travel in their explorations? We know that shortly after the year 1000, Leif Eriksson and his men sailed beyond Greenland to a place they called Vinland. Eriksson's father, Erik the Red, is often credited with the discovery of Greenland itself, not long before. It is nearly certain that Vinland was at least as far west as Newfoundland and perhaps it extended even farther south and west along the North American coast. The question of the farthest penetration by these early European explorers remains a matter of conjecture. We have confirmed that a Norse settlement existed at l'Anse aux Meadows, located at the northern tip of the island of Newfoundland, where ruins are still visible. Is there any evidence that could place them farther west, in Nova Scotia or even New England? Imagine the discovery of a structure built by these ancient Northmen in eastern Canada or Maine or even Rhode Island.

A fundamental difference remains between the long-suspected discovery of America by the Norsemen and the proven one by Columbus five centuries later. In 1492 Europe was in a condition to follow up the discoveries of Columbus with settlement, as it had not been following the adventures of Eriksson and his kin. In their time, Europe had been diverted and later exhausted by the crusades. Any early effort at exploration seemed foredoomed in the fourteenth century, one of Europe's darkest and bloodiest, topped off by the Great Plague of 1347–1349. That horror left a continent too underpopulated and broken in spirit to push its way past the ice-locked north seas to America.

In contrast, the Columbian venture followed fast upon the great Renaissance. By that time Europe had gained the science and the momentum to cross the ocean at its widest again and again, and an explosive migration followed Columbus and his fellow explorers.

But the exploits of Eriksson and his followers remain to taunt us with the possibility of an earlier discovery, no matter how isolated in history they were. Pre-Columbian maps show that Vinland may have extended over a large portion of eastern North America. Now speculation is focused on the small stone tower located near the highest point in the city of Newport, Rhode Island, and known locally as the Old Stone Mill.

The tower is cylindrical in form with a height of 28 feet and a diameter of 22 feet. Its upper portion rests on eight Romanesque arches standing on eight stone columns evenly spaced

The lower part of the tower is seen showing the columns and details of construction.

about its perimeter. The tower is constructed with stones held in place by a mortarlike substance. Many of the stones, especially around the arches, are broad and flat and mounted face to face with a keystone at the top center of each arch. The supporting columns are not mounted directly under the edge of the cylindrical mass above, but off-center just outside it so that their outer circumference is somewhat larger than that of the structure as a whole.

For over a century two theories have been put forth by their respective proponents on the time and purpose of its construction. The one most widely held has been that espoused by the noted historian and naval officer Samuel Eliot Morison, among others. Morison contends that the tower was built shortly before 1677 by Benedict Arnold, an early settler and governor of the Rhode Island colony. It was built to serve as a mill, badly needed by the local citizenry to feed their growing numbers after an earlier mill had been destroyed in a storm.

There is little question that it was used as a mill about that time, but its construction may have been much earlier. The alternate theory holds that it was constructed by Norsemen about 1350 to serve as a church and fortification during their supposed surge into the heartland of North America. In this case, it would have been adapted to serve as a mill three centuries later. The advocates of this theory of its origin link it to the Kensington Runestone, a large stone slab discovered by a farmer in northwest Minnesota in 1898. Runic inscriptions that appear on the front and one edge are alleged to describe an expedition west from Vinland in the year 1362. Ever since its discovery it has been an object of controversy. It is challenged as a forgery by some while its authenticity has been defended by others. If it is of medieval Norse origin, it (and similar artifacts discovered in other American locations) indicates a much larger and later expansion of European activity and settlement than called for by Vinland itself. The primary champion of this theory was Hjalmar R. Holand, a Norwegian-American historian active during the early years of this century. He maintained that the Newport Tower was the headquarters of these explorations and that the present structure is only a small part of the building erected for the purpose.

The two sides in the controversy have debated back and forth for years. In 1948, William S. Godfrey made some archaeological digs around the base of the tower and concluded from them, in a doctoral thesis at Harvard, that the seventeenth-century origin had been confirmed. Although he is alleged by some to have later backed off this firm conclusion, his view has represented the consensus position described later by Morison.

In 1954 Sir Arlington Mallery and two local engineers examined the tower and its foundation and claimed to have found evidence disputing Godfrey's basis for his conclusion, although not necessarily the conclusion itself. Some evidence was also found for a sighting of the tower about 1632, many years before Newport could have used it for a mill. Since then, interest in the tower and its origin has tapered off, into general agreement on its erection as a mill about 1675 by Governor Arnold. A sign next to the structure testifies to this theory of its origin. Could it have been erected around 1630, the date of its alleged first discovery? It is problematical that a mill of this size and sophistication would be built just a decade after the first pilgrims stepped off the Mayflower. Nonetheless, we cannot dismiss this possibility.

Any final resolution of this controversy must be grounded on three points. The first is the archaeological dating of the structure. Since the work of Godfrey and the three engineers, archaeology has seen many technical improvements. One of the best known is the radiocarbon dating method. This technique was first applied in 1949 and is used to date the time of death of any organic, or living, matter, plant or animal, such as bone, shell, or charcoal. This method and other newer ones can be used to date the trapped carbon dioxide in the mortar used in the joints and crevices of the tower to hold the structure and its supporting columns together. The results are inconclusive.

A second point concerns the style of the architecture. Much has been written about its construction. Was it built to be a mill or a church? Is its style that of the early Puritans or of earlier Norse design? Is its resemblance to early Norse stonework still standing in Europe of significance?

The final evidence results from a new discipline scarcely recognized in the 1940s and 1950s. This is archaeoastronomy, a science that uses computer technology to determine the degree to which a structure is aligned along astronomically significant directions. The rising and setting points along the horizon of the Sun and Moon at key times of the year are the most important of these directions.

By far the best-known object in this field of study is Stonehenge. This is the great Bronze Age neolithic circle of giant stones that has been standing for four millennia on the Salisbury Plain in southern England. We do not now know what was in the minds of the unknown race of builders as they mounted stones weighing up to 30 tons in a general circular form. They left no written record to indicate its purpose. But archaeoastronomy has progressed to the point that we have been able to discern between alignments arising from a plan and those arising only by chance. In the case of Stonehenge and many other prehistoric monuments around the world, the number of alignments toward rising

View of Stonehenge just before sunset.

and setting points of importance occur much too frequently to be a matter of chance. Today it is widely accepted by astronomers that Stonehenge served as an observatory and was even used to predict eclipses of the Moon and Sun.

The best-known alignment at Stonehenge, and perhaps anywhere in the world, is that toward the heel stone, a large, solitary stone lying about 75 meters (about 250 feet) roughly northeast of the center of the sarsen circle, the circle formed of the largest stones. Viewed from the center, the Sun rose directly behind this stone on the day of the summer solstice some 4000 years ago when the stones were erected. The obliquity, or tilt, of the axis of the Earth was then very close to 24 degrees; in the four millennia since that time, it has decreased to $23\frac{1}{2}$ degrees, but the heel stone has developed a lean, or cant, and the Sun at today's lower solstice still happens to rise just above it.

Alignments of these kinds have been found in ancient structures all over the world—in the huge pre-Inca Nazca figures on the desert plateaus of southern Peru, in the Bighorn Medicine

The heel stone as seen from the center of Stonehenge toward the northeast through upright stones of the sarsen circle. At the time of its construction, about 2000 B.C., the inclination of the Earth's axis was about 24 degrees. Now it is $23\frac{1}{2}$ degrees, and the Sun rises over the same heel stone, now settled into the soil and tilted at an angle from the vertical.

Wheel near Sheridan, Wyoming, and in similar wheels in Alberta and elsewhere. If similar alignments are found among the openings in the upper portion of the Newport Tower and confirmed beyond doubt as planned by the builders, a pre-Columbian origin is all but certain. After the work of Columbus's contemporary Copernicus, and other astronomers of his period, almanacs giving reliable positions of the Sun, Moon, planets, and bright stars were compiled. They had become widely available after the invention of the printing press at about the same time. The early Colonial settlers had no need to orient the tower to keep track of the calendar from rising times and orientations of the Sun and other celestial bodies, just as there is no such need today.

In the last few years, studies of the Newport Tower have been made in each of the three areas described here (dating the structure, analyzing its style, and uncovering its alignments). Preliminary findings suggest a possible Norse origin, but they are inconclusive and indicate that further research is fully warranted. The openings on the wall of the first level of the tower appear to be aligned with key points on the horizon that correspond to the rising and setting directions of the Sun and Moon, much as at Stonehenge. The openings are of differing sizes and shapes and appear irregular and whimsically placed as part of no apparent plan. But scientists working at the University of Rhode Island have found that the openings or windows in the tower match most of the major horizon alignments along with some others. These published results call for further investigation of the matter. Furthermore, a discovery has been made of what appears to be a runic inscription on the southwest side of the structure. When deciphered, it purports to read, "Henricus the 2nd Sunday in Advent December 10, 1116." Finally, the form of the tower closely resembles others known to be of early Scandinavian construction. A similar towerlike structure forms the central core of the Round Church built in Cambridge, England, about 1130, and still standing. Its eight interior pillars and connecting arches resemble their counterparts in the structure in Newport.

Sixteen directions form the basis for the alignment of early prehistoric monuments. Four of them correspond to the four cardinal directions—north, east, south, and west—and are easily determined. Another four coincide with the midsummer and midwinter sunrises and sunsets, the points along the horizon over which the limb of the rising Sun first appears and the setting Sun is last seen when it is farthest north and farthest south of the equator. These events, the solstices, occur on or about June 22 and December 22.

The remaining eight directions have to do with the Moon. The Moon's orbit is inclined to the ecliptic, the Sun's apparent path in the sky, by about 5 degrees and 9 minutes of arc. If it were not, but instead coincided with the ecliptic, we would witness a solar eclipse every new Moon and a lunar eclipse each time the Moon was full. But in actuality, the Moon passes north or south of the exact alignment for an eclipse most of the time.

Old Norman church of Cambridge, England. Built in the twelfth century, its interior columns are similar in design and appearance to those of the tower at Newport.

Furthermore, the orbit of the Moon wobbles about the ecliptic like a warped phonograph record, with a period of 18.61 years. It must cross the path of the Sun at two points, called nodes, and these nodes regress or move westerly along that path. Just as the Sun attains an angular distance of 23 degrees and 26 minutes of arc north and south of the celestial equator, so the Moon at times achieves an angle of 23 degrees, 26 minutes plus 5 degrees, 9 minutes, for a total angle of 28 degrees 35 minutes from the celestial equator. Just over nine years later, its angle lies at 23 degrees, 26 minutes minus 5 degrees, 9 minutes, or 18 degrees, 17 minutes. The largest possible angle is known as major standstill, and the smallest as minor standstill; both extremes are pictured in the figure on page 262. Jules Verne may have known of the standstills when he wrote his fictional account of a journey to the Moon. In his story, he placed a giant cannon near Tampa, Florida, at just 28 degrees north latitude. His scheme of firing the space

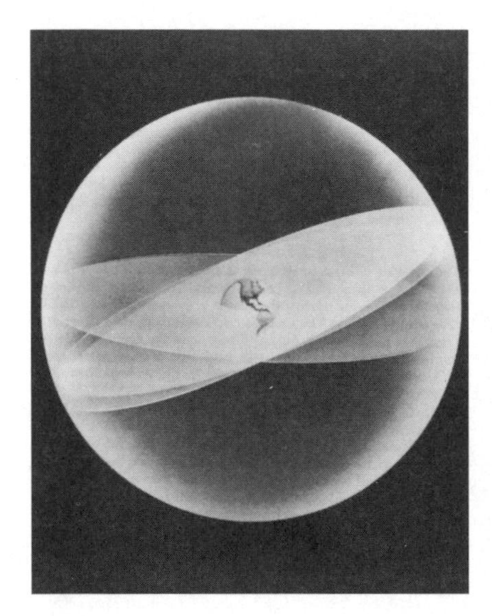

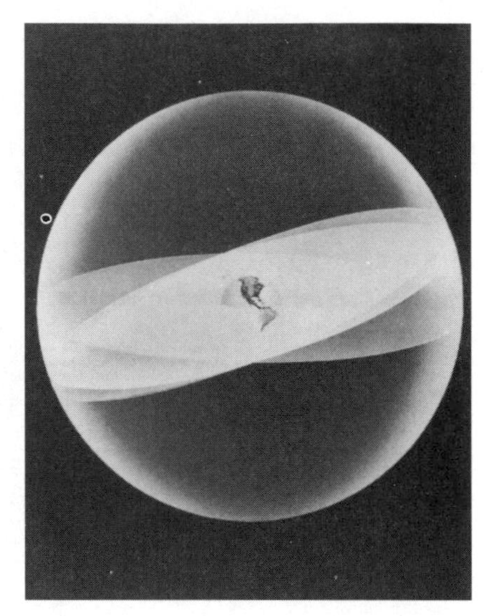

Diagram showing the Moon with its orbit at major standstill (left) and minor standstill (right). Courtesy of Griffith Observatory.

capsule directly upward is not workable under Newtonian mechanics, but had it been, his choice of Tampa made sense, since the Moon can just achieve the zenith over Tampa when it reaches major standstill.

Major standstill last occurred in 1987, and the lesser one came by in 1996. Late in the year 2005, the Moon will again be at major standstill. The variation between the two standstills is most noticeable at high latitudes. At major, the winter full Moon can be circumpolar as seen from Anchorage, while the summer full Moon does not rise at all there, and it is only barely visible from Stockholm and Saint Petersburg. Yet at minor, it can be seen throughout the full day far to the north of these cities, at the northern extremes of Alaska and Scandinavia.

Moonrise and moonset at major standstill add four points along the horizon. Two correspond to the extreme northerly distance the Moon achieves every month, and the other two to the extreme southerly points. Similarly, four others mark the north and south limits of the Moon along the horizon when it reaches minor standstill. These eight points, combined with the four for the Sun and the four cardinal points, make a total of sixteen points in all, and every one is marked by alignments incorporated into Stonehenge. Other astronomical phenomena give rise to additional alignments, including the rising and setting points of some of the brightest stars, but these sixteen are the most significant ones. They are very essential for the prediction of eclipses, solar and lunar, and for the upkeep of a viable calendar in general.

The figure on this page shows the directions to each of these points along with their names. With the exception of the cardinal points, the directions form four groups of three each. The three pointing approximately northeast are known as northern lunar major moonrise, summer solstice sunrise, and northern lunar

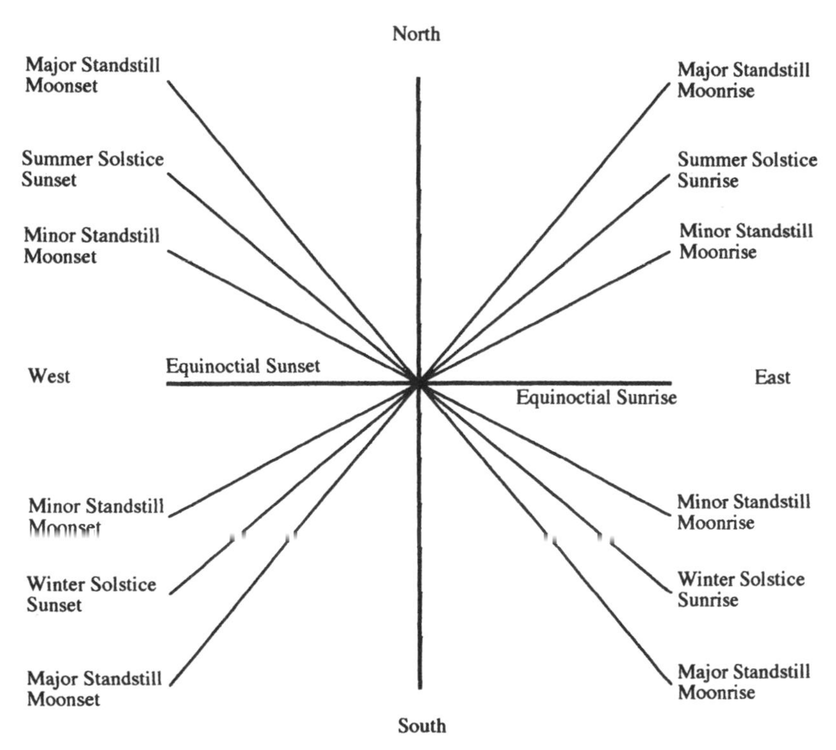

The directions of 16 alignments as oriented for the latitude of Stonehenge.

minor moonrise. Names for the points forming the other groups are shown in the diagram. This figure fits the latitude of Stonehenge, at about 51 degrees north latitude. The directions cannot be simply conveyed from one latitude to another. In each case and for each site, they must be derived anew. It is not a simple matter to calculate these directions for civilizations with no understanding of trigonometry; the points were probably gleaned from repeated observations at the appropriate times.

CHAPTER 28

New Life for Life on Other Worlds

There is grandeur in this view of life, . . .

—CHARLES DARWIN

Is anyone out there or are we alone in the universe? Two recent discoveries have brought us closer to a definitive answer to whether there has ever been life on other worlds.

For almost 400 years, astronomers have suspected that planets were not unique to our solar system. Giordano Bruno was burned at the stake in Rome in the year 1600, in part because he believed in other solar systems. He and his younger compatriot Galileo were among the first to embrace the Copernican model of the solar system.

Despite many claims made in the past, only now have discoveries of planets around other stars been confirmed. Just since 1995, at least seven new planets have been discovered, each orbiting a different star. Keep in mind that the nearest star other than the Sun is about 7000 times as distant as Pluto, the farthest planet, and that the stars with confirmed planets are as much as ten times as far away as that. Given these vast distances, we should not be surprised that even a world as big as Jupiter, our largest planet, would remain undiscovered without the sophisticated instruments that we have only recently been able to use.

The new planets are all similar to Jupiter, at least in size. Jupiter is not at all like the Earth. A gaseous giant, it has a diameter 11 times that of the Earth, and a mass over 300 times our own. It, and to a lesser extent, the other three Jovian or major planets—Saturn, Uranus, and Neptune—are all many times easier to detect across interstellar distances than is the Earth. It is not surprising, then, that all of the extrasolar planets are comparable to Jupiter in mass. We are still a long way from the discovery of an earthlike planet around any other star.

These new planets have not been *seen* in any telescope; their presence is inferred from very slow periodic motions of their primary stars. Jupiter's gravitation influences the Sun, causing the Sun to orbit around the center of gravity between the two, called the barycenter, with a period of 12 years. The barycenter is so close to the Sun (or within it) that the Sun appears to be stationary with the planets going around it. But actually the Sun goes around a barycenter between it and each planet, as Newton's law of gravitation stipulates. The result is a compound motion as it orbits the barycenter of the other planets as well. Since the Sun is a thousand times as massive as Jupiter, its orbit around their common barycenter is but a thousandth as large as Jupiter's. The Sun moves at a speed of only about 30 miles per hour around the barycenter owing to Jupiter's presence. This is slow indeed, but still measurable with our best equipment. Thus from a nearby star, our technology could just detect the existence of Jupiter from the alternating approach and recession of the Sun, but only if observed over several of those 12 years. The solar wobble due to the gravitation of the Earth has a period of one year, but the much smaller Earth causes the Sun to move with a much slower speed of just one-fifth of a mile per hour—about the speed of a lumbering tortoise.

Once planets are known to exist elsewhere, the natural next step is to determine their frequency; that is, what percentage of stars are accompanied by one or more planets. This requires a census derived from observations of many hundreds of stars, one by one, to obtain a large sample of those that do or do not share the tiny but revealing motion. The search to accomplish this aim is now under way.

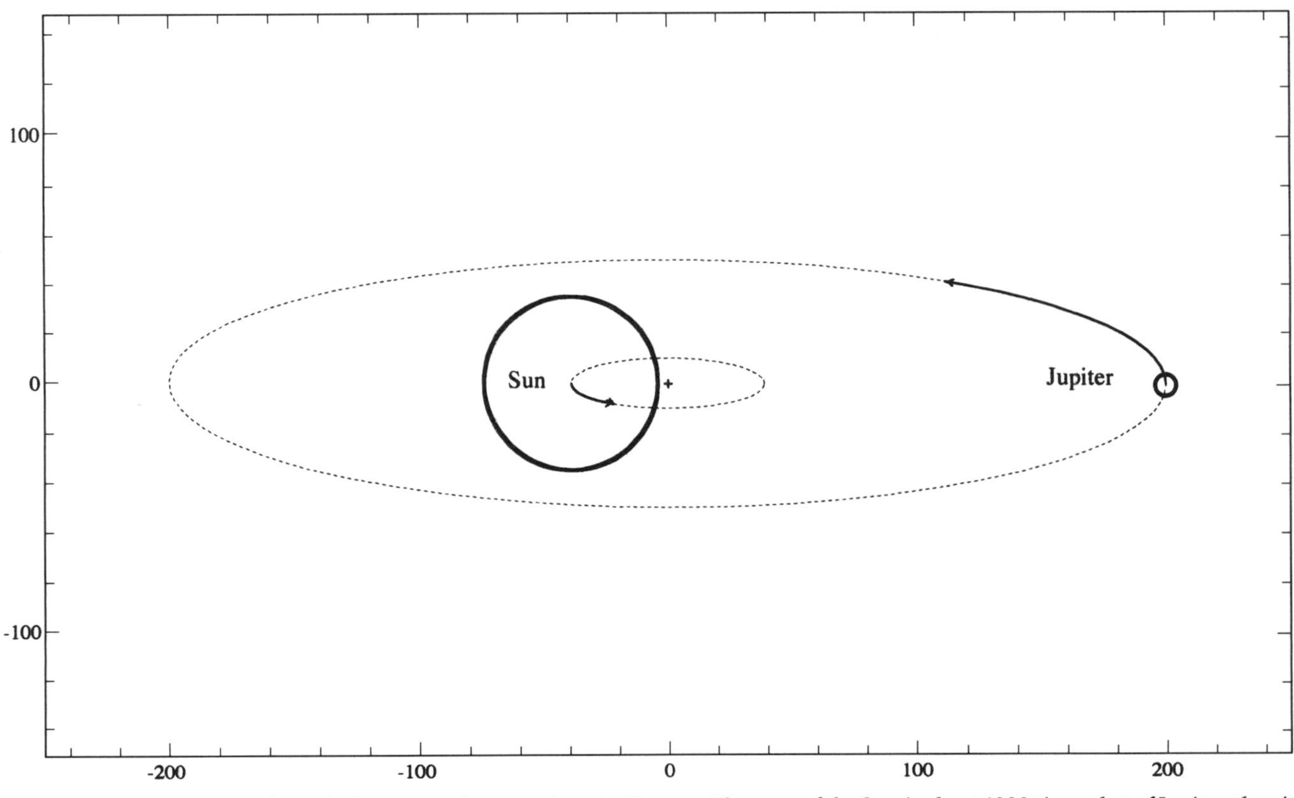

The Sun and Jupiter orbit about the barycenter between them in 12 years. The mass of the Sun is about 1000 times that of Jupiter, thus it is only one-thousandth as far from the barycenter. The Sun moves only about 30 miles per hour in its small orbit, whereas Jupiter moves one thousand times as fast.

After that, we would want to determine each of the following: the mean rate of star formation over the lifetime of our Milky Way galaxy, the number of planets per solar system with conditions suitable for the origin and evolution of life, the fraction of suitable planets on which life originates and evolves into complex forms, the fraction of life-bearing planets with intelligence, the fraction of these worlds with intelligent life that develops a technology and an interest in interstellar communication, and the mean lifetime of a technological civilization. Some awareness of each of these points is necessary to make a suitable estimate of the number and abundance of civilizations in the galaxy. Much must yet be done before we can say with any certainty that intelligent aliens exist.

A recent striking discovery has been made that life may once have existed on Mars. In a fragment blasted off of the red planet that much later fell onto the Antarctic ice cap, the remains of living things may have been found. Confirmation could be made from Martian explorations, missions now well on their way.

Clear-cut divisions cannot always be made between living matter and matter that has never been alive, and the lifelike stuff in that fragment from Mars may well fall into the gray area in between these two extremes. Life in very simple forms has been present on our world for about 4 billion of the 4.6 billion years of its existence. Any life on Mars would have formed at almost the same time as that on Earth. Exobiology, the study of life in other parts of the universe, knows of no clear concise definition of life since biologists are not in agreement as to whether some complex molecules on Earth are alive, whether they grow through metabolism and reproduce.

For the moment, we must equate life in the universe to "life as we know it," or our speculations will have little in common with scientific reality. We are aware that life on the Earth is based on the extensive combining properties of the carbon atom, and we can only conjecture about different forms of life.

Life on the Earth has been thought to be limited to two recognized forms. One, the eukaryotes, is the group of organisms with cells containing nuclei, the group that includes all plants and animals. The other group, the prokaryotes, consists of the bacteria. A recent study now reports a third kind of life existing only on the ocean floor in near-boiling water heated by undersea volcanism, with a cellular structure resembling neither of the two other better-known life-forms.

If borne out, this terrestrial discovery of another life-form would intensify interest in a confirmation of life elsewhere. It is always possible that some future unforeseen discovery could dash the hopes for life elsewhere. The possibility of relicts of Martian life will surely be treated with healthy skepticism and tested in every imaginable way. As the astronomer Carl Sagan put it, "extraordinary claims require extraordinary evidence," and such evidence may not be forthcoming. But at present, optimism is in the ascendancy.

Two other worlds in this solar system in addition to Mars may just turn out to harbor life in some form. Both of them are satellites, one of Jupiter and one of Saturn. Along with our Moon, they are among the seven largest moons in the entire solar system. All seven are larger than Pluto, the smallest planet, and three even exceed Mercury in diameter.

Europa is one of the four large moons of Jupiter discovered by Galileo in 1610. To space probes, Europa looks like an ice-skating rink wrapped around a globe. The ice is known to be water ice, and water in liquid or slushy form may be present just below the surface. Liquid water is a medium in which organic molecules may be found and life may originate. The heat that would allow water on Europa to remain in a liquid state comes not directly from the Sun, but from the enormous and varying tides raised by mighty Jupiter nearby. The tides act to deform the satellite like a grapefruit squeezed for ripeness, and the extra energy is transformed into heat in its interior.

The other candidate for harboring life is Titan, the one big satellite of Saturn, and the only moon anywhere with a thick atmosphere. The atmosphere of Titan, like that of the Earth, consists mostly of nitrogen, and the air pressure at Titan's surface is a little greater than it is here on the surface of the Earth. After the Voyager space probes verified this fact, it was thought that a greenhouse effect on Titan may have warmed its surface well above the rest of the frigid outer solar system. But recent information suggests a typical surface temperature near 300 degrees below zero Fahrenheit. If so, Titan gets demoted well below Mars and Europa as a possible living world.

With several hundred billion stars in the galaxy, and as many billions of galaxies in the observable universe, chances for life somewhere seem nearly incontrovertible. Life just must be around somewhere in all that lot. The important thing in all of this would be the confirmation of an origin of life totally independent from ours on the Earth. Life formed on our planet just about as early as it could have done so, implying that its origin might be a common development. If the same process brought about life on Mars or anywhere else, the two independent events could be seen as part of a natural process in the universe. The next few years promise to be an exciting period in exobiology and astronomy. Three new missions to Mars landed there in 1997 and probed its surface in detail. In July of that year the NASA Pathfinder probe reached the Martian surface, and its little six-wheeled companion Sojourner has delighted millions with its travels from one rock to another.

The discoveries of extrasolar planets and the possibility of life on Mars have fueled our ever-hopeful fires and evoke enormous interest everywhere. Recently it has become fashionable to assert that science is approaching an end, with only a few details still to be resolved. The case for science as a finished process is about as credible today as it was a century ago, shortly before the revolutionary discoveries of relativity by Albert Einstein and of quantum physics by Max Planck, Niels Bohr, and other giants of modern physics. The confirmation of life on another planet would likely set off a new golden age in exobiology and space exploration.

On a dark night we look at the stars and the Milky Way and we cannot help but wonder at the purpose of it all. Surely it cannot all just be for the inhabitants of our tiny planet to observe and ponder. A universe teeming with life is our only refuge from an empty cosmos, a universe that otherwise would intimidate us, even frighten us, by its very expanse.

The nearest intelligent beings may be thousands of light-years away from us; we might need thousands of years to pass a single message between our civilization and theirs. We might never meet them face to face. But great distances in space and time will not take away a single spark from the proof that life, intelligent life, whether mammalian, reptilian, or a kind of life not yet known to us, exists beyond our own small world. This discovery will transform every facet of human thought and will be a new defining moment for our race. Science fiction, with its endless speculations, has made us both aware and not aware of the shattering impact that such a new reality would bring. Perhaps if we knew beyond a doubt that intelligence and civilizations superior to ours really did exist, we might envision them as a kind of *Homo superior*. We might even see them as a manifestation of God.

Afterword

John and his father go out to look at the stars. John sees two blue stars and a red star. His father sees a green star, a violet star, and two yellow stars. What is the total temperature of the stars seen by John and his father?

—QUOTED BY RICHARD FEYNMAN
Surely You're Joking, Mr. Feynman

When I first read this excerpt, I could only laugh. Dr. Feynman had stumbled across this problem while serving on the California State Curriculum Commission, whose responsibility it was to select the new schoolbooks to be used in that state. It had appeared as a problem in a science text written for use in secondary schools and was accompanied by a table of stellar temperatures by color. In his own book, Feynman points out the utter absurdity of this problem. No scientist would ever have reason to add together the temperatures of several stars, unless it helped to determine their *average* stellar temperature. The *total* temperature of this random lot by itself is a worthless quantity of no astronomical value.

But that is not the only problem with the question. In a sky full of stars, how can John manage to see only three while his father sees a different four? On a very hazy night the visibility could just possibly be limited to the seven brightest stars above the horizon at the time, but then both John and his father would surely see the same stars. On the kind of clear night meant for stargazing, the stars taken together form the single most impressive feature of the night sky. Try to see or take note of only three or four stars on a clear night! The author of this question seems to have no feeling for either the appearance of the sky or the scientific worth of the data. He has not considered its instructive value or the imagination of the students.

We can sympathize with Sylvia Plath, who, in *The Bell Jar*, bemoans the quantitative in astronomy. "Physics made me sick the whole time I learned it. What I couldn't stand was this shrinking everything into letters and numbers." Those who bring astronomy into the classroom, the lecture hall, and the planetarium would do well to heed her lament, and to be sensitive to that wonder that so many feel for the heavens.

Nonetheless, astronomy purports with good reason to be an exact discipline, one of the first fields of study to become so. Lord Kelvin cautioned that, "When you can measure what you are speaking about and express it in numbers, you know something about it; but when you cannot . . . you have scarcely advanced the stage of science."

Are these two statements at odds? I think not. Astronomy is best served by a happy blend of theoreticians and observers, amateurs and professionals. The beauty of an elegant theoretical proof is fully matched by the beauty of the untainted sky at night. While theoretical work is proceeding, we must also make certain that the sky is not further sullied by our own light and air pollution. Consider the great loss to the world if descriptions of the clear vault of heaven can be found only in the past, in the words and pictures of others.

George Kennan, on an expedition to the Kamchatka Peninsula and beyond in 1865–1867, described a lonely, moonless night on the silent Arctic wastes (in his book *Tent Life in Siberia*). The black sky with the bright constellations of Orion and the Pleiades sparkling high overhead was vividly interrupted by the shimmer of the northern lights above, and harbored a stillness of unearthly intensity that he described as both profound and oppressive.

Kennan realized he was watching the sky in solitude under bare and desolate conditions just as humans and humanoids had done long before him. Our sky may be gentler and less intimidating than the Arctic sky that so moved him, but it need not be the less arresting for it. A dark sky filled with stars has always been one of our most cherished sights. This wonder need not and must not fade into the baleful orange glare above our cities; let the stars continue to twinkle with the fireflies along country lanes. Those stars form the one shared legacy of all people around the world, and it is by the heavens they define that we all ultimately find our way.

Appendix I

The Brightest Stars

Most of the data used in this table come from Hipparcos, the European Space Agency astrometric satellite, whose results were made available in 1997. Distances given here are much more accurate than distances from formerly available data, also shown as sources 1 and 2, for all but the very closest stars (Sirius, Alpha Centauri, and Procyon). In the final column appear the uncertainties in the distances. The probability is 2 chances in 3 that Canopus, for example, is within 16 light years of 313; that is, a two-thirds chance that it lies between 297 and 329 light years from the solar system.

Star	Constellation	Magnitude	Distance (Light-Years)			
			Source 1[a]	Source 2[b]	Hipparcos	Error
Sirius	Canis Major	−1.4	9	9	8.6	±0.04
Canopus	Carina	−0.7	74	1170	313	16
Alpha Centauri	Centaurus	−0.3	4	4	4.3	0.01
Arcturus	Boötes	0.0	34	36	37	0.3
Vega	Lyra	0.0	25	26	25	0.1
Capella	Auriga	0.1	41	42	42	0.5
Rigel	Orion	0.1	1400	910	773[d]	148
Betelgeuse	Orion	0.4v[c]	1000	310	427[d]	92
Procyon	Canis Minor	0.4	11	11	11	0.04
Achernar	Eridanus	0.5	69	85	144	3.6
Beta Centauri	Centaurus	0.6	320	460	525	47
Altair	Aquila	0.8	16	17	17	0.1
Alpha Crucis	Crux	0.8	510	360	306	20
Aldebaran	Taurus	0.9	60	25	65	1.2
Spica	Virgo	1.0	220	260	262	18
Antares	Scorpius	1.0	522	330	604	188
Pollux	Gemini	1.1	35	36	34	0.3

Fomalhaut	Piscis Australis	1.2	22	22	25	0.2
Deneb	Cygnus	1.3	1500	1830	3230[d]	1820
Beta Crucis	Crux	1.3	460	420	352	23
Regulus	Leo	1.4	69	85	77	1.5
Adhara	Canis Major	1.5	570	490	431	32
Castor	Gemini	1.6	49	46	52	1.0
Gamma Crucis	Crux	1.6	120	88	88	1.6
Shaula	Scorpius	1.6	330	270	703	136

[a]Source 1 is taken from *The Observer's Handbook*, published by the Royal Astronomical Society of Canada, Toronto, 1996.

[b]Source 2 is taken from *A Concise Dictionary of Astronomy*, by Jacqueline Mitton, published by Oxford University Press, Oxford, 1991.

[c]v indicates that this star varies in brightness between 0.1 at its brightest and about 0.8 at its faintest.

[d]The three most distant stars (Rigel, Betelgeuse, and Deneb) are too far to be measured with precision using the trigonometric parallax method. Their distances are estimated from other evidence to be about 1400, 1000, and 1500 light years away, respectively. For the others this is a remarkable degree of accuracy compared to any prior data, examples of which appear here as sources 1 and 2. Such stars as Canopus, Beta Centauri, Alpha Crucis, Spica, Beta Crucis, and Adhara have well-determined distances for the first time. The intrinsic luminosities of these and similar stars can only be found from reliable distances, and thus they are now also well known.

Appendix II
The Telescope

The telescope is believed to have been invented in 1608 by Hans Lippershey, a Dutch spectacle-maker. It was a simple affair, consisting of two lenses of different focal lengths, giving an inverted image. Galileo heard of it in the following year, when the news of its existence, and even examples of it, was spreading throughout Europe. He was only one of several who used the spyglass to view the Moon and planets, but by November 1609, he had made many improvements and had one that magnified twenty times, not just three to four times as those others had made. As a result he was the first to report craters on the Moon, the phases of Venus, the four large moons of Jupiter, and much else. Galileo was quickly followed by others, including Johannes Kepler, but his prompt publication of these and other discoveries in *Sidereus Nuncius* (or *The Sidereal Messenger*), assured him of full credit for them.

Improvements to the telescope came rapidly. In 1655, Huygens was able to discover the true nature of the rings of Saturn, and by 1671, Isaac Newton had developed the first reflecting telescope, still on display at Trinity College in Cambridge, England. This second type of telescope uses mirrors, as opposed to the refracting telescope, which uses lenses.

Throughout the telescopic era, from 1609 to the present, telescopes steadily improved in size and optical quality. By the mid-nineteenth century, the first detectors other than the human eye were developed, beginning with photography and the spectrograph. By 1950, the photocell was sophisticated to the degree that very precise brightnesses of stars were available in large numbers. Improvements in the technology of ancillary equipment have continued lately at a faster pace than improvements in telescopes themselves. The charge-coupled device, or CCD, was first used in 1975, and is today the most widely used of detectors. CCDs are of such sophistication that even the larger telescopes of amateur astronomers can become research instruments, and now well over 90 percent of all observations made at the large mountaintop observatories employ CCDs. Coupled with complex image reduction programs, such as the Image Reduction and Analysis Facility (IRAF), CCDs have increased image detection and resolution by many times over conventional equipment. Astronomy, much as any other technical enterprise, is now filled with a litany of acronyms, many known only to the practicing researcher in a restricted field of specialization.

The size of a telescope is customarily denoted by the diameter of the objective, the largest optical component (lens or mirror), in inches, meters, or centimeters. Many are named in honor or memory of someone; no matter, astronomers always refer to them by aperture size—thus "the six-inch" or the "one-meter." The power of a telescope is expressed by no less than three different quantities. One is magnifying power, set by the ratio of the focal length of the largest, or objective, lens or mirror to that of the eyepiece. A common six-inch reflector, a popular telescope for amateur viewers, may have a focal length of 50 inches. Used with an eyepiece of one-inch focal length, it rates 50 power. But change to a one-half-inch eyepiece, and the power becomes 100. Most telescopes are sold with a set of several different eyepieces to provide a range in magnifying power.

Light-gathering power is measured by the faintest star or object visible in the field of view. It depends mainly on the square of the diameter of the objective. A twelve-inch telescope, being twice the diameter of a six-inch, can see a star only one-fourth as bright. Finally, the resolving power depends on the inverse of the size of the objective. This power is defined by the ability to form distinguishable images of objects separated by small distances, such as resolving a close double star into its two separate components.

Appendix III
Suggestions for Further Reading

A Concise Dictionary of Astronomy, by Jacqueline Mitton, Oxford Press, Oxford, 1991. A book of complete definitions to most astronomical concepts and objects of interest to amateur and professional astronomers.

A Field Guide to the Stars and Planets, Second Edition, by Donald Menzel and Jay Pasachoff, Houghton-Mifflin, Boston, 1983. Full descriptions of the Sun and solar system, stars and constellations, with emphasis on the techniques of observing with binoculars and small telescopes and the objects to be seen.

National Audubon Society Field Guide to the Night Sky, by Mark R. Chartrand and Will Tirion, Alfred A. Knopf, New York, 1991, 1997. A guide to all of the constellations and to stars and planets, with the naked eye and small telescope.

Star Names, by Richard Allen, Dover, New York, 1963. A full explanation of the origins of names of stars and constellations throughout recorded history.

The Copernican Revolution, by Thomas S. Kuhn, Harvard Press, Cambridge, 1957. A readable history of the development of our understanding of the planetary system from ancient Greece to Copernicus, Galileo, and Kepler.

The Nature of Light and Color in the Open Air, by M. Minnaert, Dover, New York, 1954. This classic work is an account of sky phenomena by day and night. It explains mirages, haloes around the Moon and Sun, intensity of sunlight and starlight, and many other observed features.

What if the Moon Didn't Exist?, by Neil F. Comins, Harper Perennial, New York, 1993. How would the Earth be different if it had no moon, or was smaller than it is, or the Sun was larger? Delightful speculation on a series of "what if's" in the solar system.

Acknowledgments

Portions of Chapter 11 have been previously published in *Amicus* magazine and Chapter 15 in *Mercury* magazine. This book would not have been possible without the help of many people. To John Wareham and John Lee, my gratitude for their patience and care in providing the figures and sky charts. To Linda Shettleworth, my thanks for her tireless efforts in preparing the text and copies. I also wish to thank Joseph Caruso and Jurgen Stock for their critical reading of the manuscript, William Penhallow for providing much information in Chapter 27, and Elizabeth Lutyens for her help in editing the text in detail. I am very grateful to Charles Cutler for his advice and encouragement and to Joan Jurale, Erhart Konerding, and Edmund Rubacha of the Olin Library Reference Center at Wesleyan University, who quickly provided extensive reference material. I am indebted to my agent, Sally Brady, and my editor, Erika Goldman. Their patience and forbearance have been remarkable. Any errors in the text or the illustrations are mine alone.

Index